Hysterectomies & You

What you need to know that no one tells you

~~~~~

Pamela Clayfield, RN

Hysterectomies & You… what you need to
know that no one tells you
by Pamela Clayfield, RN

Copyright © 2018

ISBN: 978-1-387-95513-8

Online editions may also be available for this title.

To the family and loved ones who surrounded

me and gave me support when I needed it

most...

Contents

Introduction

On November 30, 2016 I had a regular pap and left the office thinking "see you in three years."

Eight days later my whole world started to tilt and another eight days after that it was upside down.

One year, less two weeks, later I had a hysterectomy to all but rid myself of a threat that I know I would have carried with me and stress over for years to come if I didn't.

Being an RN made the entire process easier for me but it also made me realize that there is a huge gap for a lay person as far as information about the surgery and recovery goes.

So welcome.

You have likely just had a discussion with your gynecologist, or perhaps your family doctor, about a hysterectomy for a medical reason that is deemed threatening enough that it requires removal.

3

Hysterectomies & You

I have been where you are at this moment. I first had the discussion with my family doctor who made the suggestion. I then saw my gynecologist and I had the discussion with him about hysterectomy.

And I've had the surgery.

I, too, spent time asking myself if I was making the right choice addressing it with my surgeon. Despite all the confidence I had going into that consult appointment my subconscious was still asking if it was what I really wanted to do as I sat in the exam room waiting for him to come in.

Being a health care professional, I've seen the big picture and didn't want to end up there. That helped a lot and made making the decision easier. With everything I had gone through for almost a year leading up to this consult I realized how much my knowledge made a difference through it all. It also made me realize that there are *more* women facing this decision who don't have training and who deserve to understand what they are going to experience having a hysterectomy.

As an RN I thought I had a really good idea of what to expect which helped make it easier to say *yes*. But as I sat in that exam room for that consult nervous about speaking up about what I wanted I wondered 1. How many things I still didn't know, and 2. If my surgeon was able to tell me everything and answer all my questions.

I wasn't expecting all the surprises I got like pain, swelling, infections, etc.

There's a *lot* of information out there and not all of it is accurate.

I decided to write this book to help you get answers to your questions and possibly help you make the decision or, at the very least, help you know it's the right thing to do.

Pamela Clayfield

There are plenty of websites that say a hysterectomy isn't the answer and that often makes the decision even more difficult to make for some people because they don't know what to believe. I'm not writing this book to deter anyone. I am simply providing the facts to you and I want you to know what's out the other side of the surgery so you are *not* surprised like I was. You have had the serious discussion with your gynecologist. He has recommended it based on his professional opinion of your issues. I'm counting on the fact that, for one reason or another, this is a procedure that has been deemed necessary by one or both of you.

Surgery is scary no matter the reason it's being done or the way it's being done. Whether it's the simple removal of a mole or open-heart surgery or a double lung transplant it's scary. There are risks involved with any procedure or surgery. These risks are significantly lower than they were years ago. There have been safety checklists and protective measures put in place.

Everyone's situation is completely different. We each have different reasons for having the surgery and our bodies react differently to surgery as well as how we heal so throughout this book I will share my story, my experience and what I learned on my journey with you so you can make an informed decision and the surprises won't be surprises at all.

Right now, think about the reason(s) why you picked up this book.

Is it to simply find out what hysterectomy is?

Do you need to know the reasons a hysterectomy is performed?

Or did you pick it up so you know what to expect before, during and after your surgery?

Hysterectomies & You

Perhaps you picked it up for a friend or your mother or your sister or significant other because they need to have surgery and have questions and concerns.

Maybe it's you who has questions or concerns about one of your loved ones.

My hope is that this book helps you and your support team get through whatever stage you might be at.

My last bit of advice for this section is: don't be afraid to ask questions. Yes it can be very uncomfortable but it's *your* health and it's *your* body. Show up with a list of questions and concerns. Since I understand that asking *can* be uncomfortable I hope I can successfully address all of your concerns here.

Chapter 1... Starting with the basics:

What is Hysterectomy?

The best place to start is at the beginning with what a hysterectomy is. It's very probably that you've heard about hysterectomy and know someone who's had one.

A hysterectomy, by definition, is the surgical removal of only the uterus and cervix and is referred to by most professionals as a Total Hysterectomy. On the rare occasion that the cervix is left behind the surgery is then called Subtotal or Partial Hysterectomy. IF, in the presence of cancer, the top part of the vagina is removed in addition to the uterus and cervix it is referred to as Radical Hysterectomy.

The reason for your surgery, which will be covered in Chapter 2, will dictate if your ovaries will be removed or not. In the presence of cancer they are automatically removed. The surgical removal of an ovary is referred to, in medical terms, as Oophorectomy and when the fallopian tubes are taken it is called Salpingectomy. Sometimes only one ovary and/or tube needs to be removed like with an ectopic (tubal) pregnancy or an ovarian cyst. Because of the advances made in genetic counselling and how available it now is, a woman can find out if she carries one of the two positive genes for breast and ovarian cancer. The program helps her reach the decision to have a Bilateral Oophorectomy if she wants to eliminate the risk.

If everything is being removed it's called a Total Hysterectomy with Bilateral Salping-oophorectomy. In the presence of cancer they may also take lymph nodes which sit very close to the reproductive organs.

Those are the simplest definitions and normally the reason for your surgery would define what you will have removed, but this is, unfortunately, in theory. So we know if the reason is cancer, uterine or ovarian, everything will be removed. However, if you have fibroids then a subtotal would suffice and be scheduled. Whatever you may discuss, Subtotal or Total Hysterectomy with your surgeon, rest assured, he will go on to explain that even though that's the plan, once he gets in there he's going to take a close look at all the structures and take anything he feels necessary. Therefore you could lose one or both ovaries anyway.

The route of the surgery is the next thing you'll discuss. There are only three ways of doing the surgery and these are as follows…

Abdominal hysterectomy… a traditional incision is made low in the pelvic wall just above the pubic bone; it runs just above or at the hairline. The recovery time is the longest with this procedure at 6-8 weeks.

Pamela Clayfield

Vaginal hysterectomy… the surgery is performed using a speculum in the vagina to gain access to the organs and cuts are made to remove what is necessary. The vaginal hysterectomy can be done with the assistance of a laparoscope and would then be referred to as Laparoscopic-assisted Vaginal Hysterectomy. The recovery time from this is indicated at about two weeks.

Total Laparoscopic Hysterectomy… this uses small incisions through the abdominal wall through which the surgeon visualizes the structures and uses small tools to remove the tissue. The uterus is cut into tiny pieces for removal. The recovery time is indicated at two weeks or less. Not many surgeons perform this procedure because it's so new, only being introduced in 2005.

Your body needs to be conducive to what you can and cannot have done… if you've never had a vaginal birth or only one you probably won't have the "descent" required for the surgery to be done vaginally. This means that when you push down, like when you have a bowel movement, the uterus doesn't push far enough into the vagina to allow for the surgery to be done vaginally.

There are only so many surgeons who are trained to do the surgery laparoscopically so if you insist on your hysterectomy being done this way your surgeon can refer you to someone who does it but there are no guarantees on how soon you will get a consult with the new doctor or how long you will wait for the surgery with that doctor. In the instance of cancer, for example, time is of the essence. Also in the presence of cancer you might consider having it done

Hysterectomies & You

abdominally anyway to allow the surgeon to look around inside the abdominal cavity for other bad cells.

Next, we put it all together. If you're just having your uterus and cervix removed through an abdominal incision then it is referred to as a Total Abdominal Hysterectomy. Vaginally would be a Total Vaginal Hysterectomy and so on.

Chapter 2... Reasons for

Hysterectomy

When it comes to hysterectomies there are a million reasons why one would be done. Well, not really a million but there's definitely a long list of reasons why hysterectomy would be recommended. The most common are as follows...

1. **Fibroids** — these are benign (non-cancerous) tumours that take years to develop and are made up of cells of the uterine lining. They do grow and can cause very painful, heavy periods especially when you're going through menopause.
2. **Cancer** — the ugliness of malignancies is the biggest reason for a hysterectomy to be done; and to be done quickly. This will get you a very fast surgical consult and O.R. date. Cancer is essentially the abnormal growth of normal cells

where the DNA has been damaged. This causes the cells to mutate — change the structure of the cell. These cells tend to duplicate at a far faster rate than normal, healthy cells. Reproductive cancers can be uterine, cervical or ovarian.

Uterine, or endometrial, cancer is the abnormal growth of previously healthy uterine cells that form a mass/tumour. Like most cancers the actual causes are unknown but the risk increases with obesity, never having a child, periods that started before the age of 12 or not reaching menopause until after 55, estrogen therapy, family history or taking tamoxifen or having radiation treatment to the pelvis for other reasons such as colon cancer.

Cervical cancer occurs when normal cells, also known as squamous cells, on the cervix start to grow abnormally and out of control. It is usually found at a very early stage through regular pap tests. Most cervical cancers are caused by a virus called Human Papillomavirus (HPV) which is transmitted through sexual contact with someone who has it. There are many different strains and most of the time the body will be able to rid itself of the virus like it does a cold but sometimes it can't and the virus damages the DNA and the cells start to change shape and size which is referred to as Dysplasia. These cells continue to mutate and eventually become malignant. Though most are caused by HPV, there is also a small chance that it forms on its own.

Ovarian cancer develops when normal ovarian cells start to mutate. The risks are similar to those of Uterine cancer — obesity, never having a child, periods that started before the age of 12 or not reaching menopause until after 55, estrogen therapy after menopause, and, of course, family history. There are factors that also decrease risk and those include birth control pills, tubal ligation and breast feeding.

3. **Uterine Prolapse** — this is the sliding of the uterus from its normal position into the vaginal canal. It can cause a great deal of discomfort as well as repeated bladder infections as it puts pressure on the bladder trapping urine therefore not allowing it to empty completely which creates an environment for bacteria to grow and multiply. If uterine prolapse is not treated it will continue to prolapsed out the vagina and put pressure on surrounding tissue including the bladder and colon. This could cause permanent damage to these structures. Even waiting for a surgical date could cause irreparable damage.

4. **Menorrhagia and dysmenorrhea** — these are usually painful and heavy periods most often during menopause and more often than not related to fibroids.

5. **Ovarian Cysts** — every woman gets simple cysts on her ovaries during ovulation and these cysts should simply disappear as the menstrual cycle completes. Usually they are painless and cause no other symptoms. However, some women are unfortunate to develop Polycystic Ovarian Syndrome (PCOS) where the ovaries contain large numbers of cysts which can cause the ovaries to enlarge and, eventually, cause infertility because the cysts severely interfere with ovulation. The enlarging ovaries put pressure on surrounding tissue causing pain.

6. **Endometriosis** — this is another of the sneaky issues a woman deals with that is incredibly difficult (okay, nearly impossible) to diagnose in a traditional sense using blood tests and imaging. It's often diagnosed after everything else has been ruled out. The fallopian tubes don't connect directly to the ovaries so during a woman's period the tissue that breaks off that we usually collect in pads or tampons can also travel the other way and be released by the fallopian tubes into the abdominal cavity where it adheres (sticks) to any surface it can find… organs, intestines, outside of the uterus, ovaries, bladder or lining of the abdominal cavity. It continues to respond the same way as it did inside the uterus. It grows the extra layer and it bleeds. That tissue will also cramp and any tissue it's attached itself to will respond in the same way. This tissue doesn't show up on ultrasound. Most often a laparoscopy to

Pamela Clayfield

have a look inside is the only way to get a true diagnosis.

Multiple other reasons include chronic pelvic pain which is often related to endometriosis; adenomyosis which occurs when the uterine lining thickens; and abnormal vaginal bleeding, especially after menopause.

No matter the reason for it, hysterectomy has obviously come up in conversation for you to be reading this book.

Chapter 3... Risks—What are the risks of hysterectomy?

There are always risks involved with any procedure or surgery and the risks are not just related to the actual procedure itself. Hysterectomy is considered major abdominal surgery no matter how it's being done. You're usually put under a general anesthetic, incisions pass through multiple layers of tissue, an organ is being removed, your other organs don't like to be touched, there are multiple blood vessels that have to be cauterized and it takes time to heal.

There are always risks with a general anesthetic including the rare reaction to the medications used. If there are any cardiac or respiratory health concerns an epidural may be offered as a safer alternative but that comes with risks of its own including the rare chance of permanent spinal damage. Other risks include:

Pamela Clayfield

Infections:

Surgical site(s)/wound(s)... Antibiotics are given through your IV during the surgery to help prevent infection within the tissues affected by the surgery. You may also receive antibiotics post-surgically as well. Sometimes this depends on the reason for the surgery and sometimes it depends on what they find and what they have to do when they get in there.

Urinary Tract Infection (UTI)... You had a catheter inserted and though it's inserted under sterile technique even that's not always perfect.

Vaginal infection... Deep inside, you have an incision line where the cervix has been removed from the upper end of vagina and it has to be closed. Because it's warm, dark and moist it's where bacteria thrive.

Lung infection... Being in hospital exposes a person to all kinds of bugs and your immune system is decreased due to surgery plus you're not as active which can encourage these bugs to settle in the dark, warm, moist environment of your lungs.

I will cover off symptoms and treatment in later chapters.

Other risks:

Excessive bleeding… Even though they do everything right during surgery once in a while there is a problem. They could nick something or they could have a hard time getting bleeding to stop. They do their best to stop the bleeding but if they can't they may have to resort to blood transfusion. Any bleeding complications during surgery increases the chance of bleeding after surgery not only internal but external. This is why everyone keeps checking your dressing after your surgery… they are checking for pooled blood.

Blood clot… Because of the surgery itself as well as the anesthetic followed by decreased movement that comes with being in bed post-operatively, there is a chance of a blood clot forming usually in the calf of the leg. This is called Deep Vein Thrombosis (DVT). Unfortunately these can break off and travel to the lungs where they get stuck in the smaller vessels. This is called a pulmonary embolus and can be deadly. This is why they are now using the preventative measures of giving you blood thinners to try to all but eliminate this possibility.
Blood clots can also form around the wound both internally and externally which, again, can react similarly to those of a DVT.

Bowel adhesions… Probably one of the most common complications of abdominal surgery they are caused by scar tissue, which gets very sticky when forming, which causes other abdominal structures, such as the bowel,

Pamela Clayfield

to adhere, or stick, to other structures. This is one of the reasons why it's incredibly important for you to pass gas while still in hospital.

Ongoing bowel issues… After the uterus has been removed, there is a gaping hole in your pelvis. The surgeons fill this gap with some of your small intestine. This can cause the large intestine to collapse into that area which can put pressure on the healing tissue at first but can also create a bit of turmoil as far as passing stool is concerned which can slow down the passage and cause constipation. I will talk more about constipation in later chapters as well.

Most of these risks should be covered by your surgeon at the time he suggests hysterectomy because it's important to be aware. Of course most of these complications are rare but it's still important to cover all of them.

Chapter 4... Making the Decision

We all have to make decisions in life, we make them every day. There are the small, easy ones… what peanut butter to buy, the route to take to work or what time to set the alarm. Making big decisions in life is always a challenge because it involves more than just us… it impacts others too. It doesn't matter if it's buying a car or new appliance, taking that job promotion, what to name your child or taking your doctor's advice.

Most people take the advice of their doctor, rarely ask any questions, rarely advocate for themselves, and carry on. For those people, the answer to having a hysterectomy will be an easy *yes*. They will simply say *okay*. I have had patients come back to the office saying the surgeon refused to do a hysterectomy and that's all they wanted. Other patients have come back saying all that was offered was surgery and they're not ready to make that decision yet. I can guarantee that you will still ask yourself, at some point, if you made the right decision. We all do that when we're

Pamela Clayfield

scared. It's the fight or flight response and we question our own sanity and ability to make decisions in those moments.

For those of you with a new cancer diagnosis there is often only one option, sometimes no option other than treatment, and very little time to think about surgery if it's an option, make the decision and be at peace with it. You will have your surgery booked as soon as possible followed by a consult with an oncologist to discuss further treatment options once the pathology report has been received and the type of cancer has been identified. Many surgeons will send you to a tertiary care centre so that gynecologists with better equipment and even more experience can do your surgery.

For those of you who have had problems for years with periods or chronic pain, you saw your gynecologist in consult and very likely had hopes of a hysterectomy being offered tucked in the back of your mind because you would do *anything* to get relief from the horrors you experience monthly and, sometimes daily!

It's important to take time to think about the options you have been given in order to make the right decision for you. If you are still having periods you need to realize and accept that you will no longer be able to have a baby. For many women, this fact alone makes this an extremely emotional decision, especially if you're still in your 30s and this decision is because of severe endometriosis or a malignancy. Once you've had the surgery it's too late to turn back the clock.

You may also be facing a questionable diagnosis. Maybe nobody has been able to get an actual diagnosis from samples taken. Again, you may not have a choice

Do your research. Talk to your family, friends and your significant other because it impacts everyone, even if temporarily. Unless it's urgent, postpone it if you need more time. There's nothing wrong with waiting and being certain before you face any surgery.

Chapter 5... The Beginning of my Story

I went for my pap when I was due and didn't think much of it. Eight days later I got a call from the doctor's office but the message said only to call them back. I tried and tried but had no luck getting through. It was after another eight days had passed and I got an email from them saying I had been referred to a gynecologist that I used my role in family practice to call the lab and ask them to forward my results to us.

HGSIL... High Grade Squamous Intraepithelial Lesion.

GULP.

I'd had ASCUS... Atypical Squamous Cells of Undetermined Significance... many years before but I'd never had any abnormal cells after it had cleared. You may be more than familiar with some of these terms and what follows.

Now it was High Grade Squamous cells with none of the steps in between. This was not good. A colposcopy was, of course, recommended.

A colposcopy is pretty much the equivalent of a biopsy — with no freezing. The procedure doesn't take long but I didn't know much about it and was incredibly anxious.

It's similar to having a pap but in the hospital unless you're lucky enough to have a gynecologist with a set-up in his office. The difference is the surgeon looks through a scope, which stays outside the body, that allows him to see where the bad cells are. Once he's identified the area or lesion you'll be told you'll feel a pinch and you really will. This will be done at least twice before it's over. You'll be monitored for about 10 minutes after the procedure is done, given discharge instructions and a drink of water or orange juice before you're discharged and on your way home.

The results of that confirmed High Grade Squamous cells and identified them at level CIN3 (Cervical Intraepithelial Neoplasia) the level that is used interchangeably with Carcinoma in Situ (CIS)… meaning cancer cells that are residing (still) in their own little bubble.

When I saw my specialist six weeks later he said we'd have to do a LEEP/Cone which is short for Loop Electrocautery Excision Procedure and Cone Biopsy where approximately 2-3cm would be spliced from my cervix with hopes that the actual lesion and all the bad cells would be retrieved. I told him he could take everything with a hysterectomy but he said "we weren't there yet" and at the time I didn't think to argue. Looking back I wished I would have pushed.

Pamela Clayfield

I was so nervous the morning of that procedure. I was given the discharge instructions to read and a copy to take home with me before I was taken to the same room where the colposcopy had been done. After my legs were back in the stirrups, a pad was stuck to my thigh to ground the electrical current generated by the cautery instrument. My cervix was frozen which stung and burned until it started to work. After that the procedure itself was over in just a few minutes. I was transferred to a reclining chair outside the room for about 20 minutes, had my vitals checked, drank a cup of orange juice and finally was allowed to get dressed and go home.

Two weeks later the results showed up at the office and showed that there *were* actual malignant cells mixed in with normal and abnormal cells in the 1 inch chunk of tissue he had removed. There were also bad cells identified at the margins which meant some could still be on my cervix. When I saw the surgeon six weeks after the procedure he recommended another biopsy three months later. Not only was that part of the follow-up screening but it would allow him to see if any of those cells actually remained.

Between my follow-up appointment and the next biopsy I saw my own family doctor and he asked why I hadn't just had a hysterectomy since I was 42 with no plans to have any more kids.

That planted the seed!

We discussed it further and he advised me to think about it and to bring it up again at my next consult if I felt strongly about it. He also told me that if my surgeon wouldn't do it to come back and he'd find me someone who would.

Hysterectomies & You

I thought a lot about hysterectomy as an option after that. It would greatly decrease my anxiety not only now but for years to come when I wouldn't have to continually wonder if the next pap would be a positive one again and I'd have to go through it all again or find out it had become an invasive cancer and it was too late. Stopping it before it got that far sounded like the best idea.

The biopsy in August showed the LEEP/Cone had actually removed and destroyed everything but, in consult six weeks later, he still wanted to wait and do yet another biopsy in February and that's when I put on the brakes!

"I want to discuss hysterectomy as an option because my anxiety levels are through the roof. I stress about the procedure and again about the follow-up consultation. I'm (now) 43 and my daughter is 19. I'm not having any more kids so why can't we just put an end to the threat completely?" It all came tumbling out in one breath. I was afraid if I paused I'd chicken out from saying the rest.

He said he would do it which shocked me but I was relieved. It was moments after he said he'd do it that I became anxious about whether I was making the right decision.

One step at a time.

He checked me and wasn't able to do it vaginally so that meant it would have to be done abdominally. I wasn't concerned in the least about how he would do it. I also wasn't worried about the scar knowing they put it nice and low. FYI it has no impact on my ability to wear a bikini.

When I stepped out of the room after him I was shocked when he told me he'd had a cancellation for November 16. That was a month and five days away and I

Pamela Clayfield

had to get clearance from work because you're supposed to be off for six weeks.

"Pencil me in. I'll let you know ASAP." I hoped my boss, being a doctor, would understand and give me the time off.

Thankfully my boss agreed and I confirmed the date two days later... the decision was made. For the next five weeks I asked myself only once or twice if I'd made the right decision.

I hoped I had!

Chapter 6... Booking and Preparing for the Surgery

By the end of your consult, your head will be swimming with information and everything the surgeon said. You'll immediately also be given dates and possibly be expected to choose one on the spot. Try to stay focused. Let's try to break it down.

Booking

If you have been diagnosed with cancer one of two things will happen. There will be a need to move fast but if your surgeon isn't confident to take your case, usually because of the possibility that it's spread to the lymph nodes, you'll likely be referred to a tertiary care centre as mentioned in Chapter 4. Larger centers are usually attached to universities and hire specialists who have been trained specifically for these specialty surgeries due to malignancy. It's important to keep six weeks in mind no matter what procedure you're having. You still have tissue that's going to be abused, especially on the inside, and it's going to take time to heal. That includes muscles, ligaments, blood vessels and abdominal incisions as well as an incision to close the top of the vagina and create the vaginal vault as it is referred to after surgery.

Due to regions having more doctors than operating rooms surgeons have operating room days; they don't get to just pick a day at random… often the hospital creates the schedule and sends it out to the surgeons. They'll plan in three month blocks and don't send out the schedules well in advance. My surgeon told me in October that it would be January/February/March but he didn't have that schedule yet. He asked if I had a preference for one of those months because they'd put me on the waiting list and call me to book when the schedule arrived. It was after that conversation that he looked at his book and found he'd had a cancellation. I felt incredibly lucky because it turns out his office didn't get the surgical schedule until December.

It all happened very fast. I was trying to digest everything the doctor had just told me then I was given a date that I wasn't expecting! You're schedule will probably start racing through your mind. My surgery was five weeks before Christmas. When he said the date, November 16, it didn't really seem that close to Christmas. It wasn't until after I got the okay from work and confirmed with the surgeon's office that it started to sink in. Everything that I would need to try and get done — including decorating — before my surgery, in order to enjoy the holiday season just a little bit flashed into my mind. I began to realize that any kind of Christmas party or gathering would very likely be off the table. I figured I'd probably be okay, at least, for family gatherings — or so I hoped. All of this occurred to me *after* I had accepted the date. Another reason for me to try to advise others if it's not too late!

My best piece of advice for this is to sit down with your calendar in the weeks before your consultation appointment and write down *everything* that you have going on… birthdays, anniversaries, Christmas, Easter, Christenings, etc. Record them all as well as anything you may be responsible for on those occasions. I suggest you give it at least eight weeks before you're expected to cook and lift a turkey out of the oven or host a retirement party or attend that Christmas party. You may still need to decline the invitation. We'll talk more about that starting in Chapter 8 where I will begin to cover all the post-op issues they don't tell you about!

Preparing

Once you get a potential surgical date the first thing you need to do is discuss it you're your employer and confirm the time off from work — some surgeons say six weeks, some will tell you eight — and see if there will be a replacement who will do your job in your absence or if your job will be merely covered by existing staff which means work that you need to address will pile up on your desk waiting for your return. If you don't work then you're probably the luckiest of us all! I'm jealous.

When you book the surgery date with your surgeon's office they will give you a package from the hospital that will contain forms for you to fill out and general information about having surgery at their facility. You'll have to go for a pre-surgical appointment at the hospital because you will be staying at least one night if you have a vaginal or laparoscopic, at least two nights if it's an abdominal and longer if there are any complications. Essentially they want to check you out to make sure you're healthy going in. The pre-surgical appointment will act as your admission which means you will be able to bypass patient registration on the morning of your surgery. Of course every hospital is different so this will vary depending on where you live.

At your pre-surgical clinic appointment they will review the forms you filled out. The bulk of the appointment will focus on your medications and allergies which will be reviewed in detail and you'll be told what you can continue to take and what has to be stopped and when. *All medications* means *anything* you put in your mouth at any point in time... for example, ibuprofen even if you only take it once a month for a headache. You can't take it for several days before your surgery because it has blood thinning properties. The "all meds" also means they want to know all vitamins or herbals you take. Don't be surprised if some of these will be discontinued before your surgery. A multivitamin contains Vitamin E which can also act as a blood thinner so they want you to discontinue that until after your surgery.

You will, of course, also be advised what medications you *can* take up to and including the morning of your surgery. All of your prescription medications will be logged so they are ordered and administered throughout your hospital stay.

Be sure to review every bit of the document they give you *very* carefully. I can't stress this enough. The medication record they create becomes your medication record during your stay in hospital so you want it to be right. I didn't realize they had put DAILY on a medication that I take at bedtime and I didn't get it at bedtime the first night despite the label on the bottle staying "at bedtime." Unfortunately it was a drug that builds up in the body over time and if I skipped a dose I'd go through withdrawal. I certainly didn't need those symptoms that night because I had not only just had major abdominal surgery but I was also sicker than sick from the anesthetic for the remainder of that day. Lucky for me I had actually packed my own medications just in case. I had a feeling so I took what I needed from my supply. I found out about the DAILY when that medication was brought to me at breakfast the next morning and I had to explain that those are bedtime meds and that I can't take them during the day because they cause severe lightheadedness and dizziness. When I got home I looked at the record they had prepared during my pre-surgical appointment and it did, in fact, say DAILY. They had recorded it wrong.

Lastly, you need to give some serious thought to what type of pain medications you want because not everyone can tolerate what they offer which is commonly an Oxycodone product commonly known as Oxycocet, Percocet or Endocet. If, like me, you have *ever* taken even one of these and didn't like the way you reacted to it or how you felt on it then now is the time to share that and be adamant you do NOT want this drug. That pertains to any drug you have ever taken and reacted to badly. Most pain medications are narcotics and you can become addicted. If you have a history of addiction you need to share this with your doctor *and* the pre-surgical clinic staff. If you take only what you're prescribed and only when you need it, you *should* be okay.

Pamela Clayfield

Once your medications have been reviewed in detail you will also, have blood taken and they'll make you swab your own derriere—yes, you read that right—they want to make certain you don't have MRSA (Methicillin-Resistant Staphylococcus Aureus) prior to admission and surgery. MRSA is one of a few bacteria that has become immune to almost all available antibiotics and can go on for years.

Everyone will have a pre-surgical visit. In addition to the basic pre-surgical, depending on your age and health history, you may also spend some time with an anesthesiologist at this appointment who will review your pre-anesthetic questionnaire and your health history to determine whether you will be able to have a general anesthetic or if they will have to alter that plan. Cardiac issues or asthma are examples of reasons to be seen and assessed in advance by this department.

I was also told I had to stop at the hospital pharmacy on the way out to buy two antiseptic sponges to use in the shower the night before (because I shower in the evenings? NO!) and the morning of the surgery.

In all honesty, I couldn't figure this out. Nothing in my training or current nursing experience could rationalize the reason for it. I still can't. I did all my laundry *and* changed the beds the day before my surgery to prolong the time until it all needed to be done again. To wash with a special antiseptic soap then go to bed made little sense since most people will still probably sweat, at least a little bit!

At the time of my purchase of these two sponges I received a brochure with instructions which advised scrubbing with the sponge from shoulders to knees! Well my shoulders weren't going to be anywhere close to the surgical site, nor were my knees because I was having an

abdominal hysterectomy. I also had concerns about having
an allergic reaction because I am allergic to latex, bandage
glue and some detergents. I used only one of the sponges,
the morning of my surgery and I merely scrubbed the
approximate surgical field… belly button to pubic bone, hip
to hip. That was it. I used my own soap everywhere else.
And guess what… I had no infection at the surgical site!

Other Preparations

The pre-surgical hospital appointment is done within
three weeks of your surgery. If you haven't already started
planning, this is definitely the time you need to start.
Prepare a list of chores that will need to be done and, assign
the tasks to those who will be helping you. Who will watch
the kids, the dog, the cat during and after your hospital stay
and how will you get to and from the hospital. You will not
be able to do housework until your post-operative follow-up
with the surgeon which will be 6-8 weeks after your surgery.
Clean your house, do your laundry, put things away, change
your beds and go shopping so your cupboards are stocked
up. Talk to your family and friends about helping you do all
these chores and errands while you can't. If it will help,
make a schedule. You will need people to shop for you, keep
the house clean, do your laundry for you and, for a short
time, possibly even cook meals for you.

Knowing what chores and tasks need to be done and
having an idea of who will do them will create less worry for
you especially after your surgery when you get home and
are feeling overwhelmed and useless. I didn't create a
schedule for things to get done and had to keep texting my
mom to see when she was coming over because I needed

Pamela Clayfield

laundry done or I was running out of milk. Plan ahead for meals because you won't have someone bringing you a tray three times a day. If you'll be on your own for any reason... your spouse works, you don't have a spouse, etc. you could cook some large meals ahead of time and freeze portions so all you have to do is microwave it for a meal. You could do this with several different casseroles and have a variety.

You need to take time out of all this to prepare yourself emotionally for the surgery... you may be at peace with it but you don't want to wake up in recovery, realize it's done and have twinges of regret. You will still have moments after anyway but preparing is key.

For some people, acceptance is easy and straightforward. I was 43, I had a daughter almost 20 and I had decided many years before that I was not having any more children. It was more important for me to be around for her than to save my uterus for no reason. I still had my moments but mostly I was okay. I was looking forward to no longer having painful periods which came as an added bonus to all this.

For others, as mentioned in Chapter 4, it will be incredibly emotional. There might be a feeling of loss... losing a piece of yourself, and the piece that gave you the ability to be a mother, may make you feel like you're no longer a woman. It's important to come to terms with this decision being the right one and accepting it before the day of surgery. You will still be a woman. You will gain some freedoms if you still had periods and some say they feel freer as far as sex goes as well. It will only change you if you let it change you. But take the time because acceptance is still important prior to surgery.

Hysterectomies & You

Stock up on activities you like to do. I felt lucky that my surgery fell during the holiday season because I was able to watch hours and hours of sappy TV Christmas movies. I also bought a Christmas puzzle book to work through and I stocked up on books to read. I had planned on getting back to my writing but I wasn't there yet... I was still suffering writer's block from everything I had endured over the course of the year.

You also need to plan and pack your bag. Outside of a CPAP unit if you use one, glasses with a labeled case and dentures the hospital doesn't provide you with any guidelines for what to bring with you. There are a few items that will help you feel more comfortable and I ended up taking two bags! It was worth it to be able to create this list for you. My first suggestion is finding something comfortable to sleep in. It makes you feel so much better when you can ditch that horrible hospital gown! If you wear nightgowns find one or two you really like and make sure it's not see-through. Tied in first with your own sleepwear... underwear because they give you mesh underwear to hold the pads in when you get up to your room.

Definitely take your own supply of pads because they have the thickest pads ever made. They order one size and the size they order is used on the maternity floor and you really won't need that much padding. Just remember, no tampons. If you're post-menopausal you will still need pads because you will still have vaginal bleeding and discharge.

Anything that will help pass the time is crucial because even though resting and sleeping is important you won't sleep the entire stay. Of course you'll likely have visitors and you will most likely have your cell phone with games, email, Facebook and texting but you'll get tired of that too. I had a magazine and a book of writing exercises. I wished I'd had more magazines because the articles are short and easy to read.

Pack your personal toiletries but don't get too excited about showering. Chances are slim they'll allow you to do that, especially if you have an abdominal incision. When I asked my surgeon about showering at my pre-op appointment with him he said I'd be able to shower *when I got home* which told me I wasn't going to be showering on the unit. But you will want your own face wash and moisturizer, toothbrush and toothpaste, a hairbrush or comb, lotion and, possibly, your make-up for when you go home. I also took my essential oils which were not only helpful but a huge hit with my day nurse!

You will need to take a glass case for your glasses if you wear them and you will need to have that with you when you are admitted the morning of your surgery unless, of course, your glasses are for reading and you can see everything else around you without them. If you wear contacts you will still need to take your glasses as you won't be permitted to wear contacts during surgery. You can start wearing them again the day after so bring them and lens care products if required. The important thing for me was getting some *normal* back the day after surgery and I recommend it.

You may also want to tuck away a few days of meds in a small pill box somewhere just in case there was an error in your record and something is missed, as previously mentioned.

Last but not least, take your own slippers because those fake slipper socks are horrible and so are the old foam ones they used to give you! Make sure they have rubber bottoms to prevent slipping!

Hysterectomies & You

If you're having an abdominal hysterectomy another suggestion is to go to your local health supply store and pick up some sort of dressing for your incision for when you're at home. This is mostly just protection for the first few showers you'll take once you're home. Your surgeon will likely tell you that you can leave it open once you go home if you choose. I recommend Tegaderm +Pad. Pick up a few that are 3 ½ x 9 inch (9 x 25 cm). These are a clear film dressing with a small strip of non-adhesive dressing in the center which fits perfectly over your incision. If you're having a laparoscopic surgery you'll probably be able to just cover your small incisions with Band-Aids. I was only concerned about the water spraying accidentally on my incision that made me go looking for something to cover it with for the first week. You can, of course, leave those at home as the nurses will change your dressing while in hospital.

If you still have periods your gynecologist will order a urine pregnancy test that you will have to do about three or four days before your surgery so they are certain you're not pregnant. It didn't matter that my significant other had a vasectomy years ago. It's standard procedure in case of immaculate conception I suppose.

Lastly, they ask that you do not shave around the surgical site(s) in the days leading up to the day of surgery as it can increase the chance of infection.

Chapter 7... Surgery Day

If you're anything like me, you've been stressing for days, probably longer. Your sleep has been interrupted, to say the least, and yet the time has passed anyway and here you are. You may have managed to sleep with no problem or you may have been awake all night nervous about the surgery and wondering if you're prepared and hoping you haven't forgotten anything. Despite what everyone thought I wasn't concerned about the surgery itself. It was the recovery that worried me. How would I feel? How would my pain be? What kind of impact would it have on my back condition? How would I manage? I only have one bathroom in the house and it's upstairs… would I be able to do the stairs? These thoughts and worries had been with me for weeks and had impacted my sleep. I didn't sleep great the night before my surgery but I did sleep more than I thought I would! I thought I'd be awake all night tossing and turning. Part of the pre-op instructions I had been given by the hospital was to get up at 5:00 a.m. to drink a cup (250 ml) of apple juice. This fairly new practice is supposed to prevent nausea and vomiting post-operatively. If you received similar instructions your night will be cut short at 5:00 a.m. unless your surgery is in the afternoon. No matter what you'll probably be tired but that's okay... you'll be having a sleep!

Unless you are able to get back to sleep after you have your apple juice you'll likely end up watching the clock, counting down the minutes until your alarm goes off and you have to get up and shower... using that special sponge! Make sure you leave a moment or two to take a few deep breaths before you head out the door. Hopefully the person taking you will be able to stay at the hospital with you and be there for you at least until you're in your room after surgery. You almost have to have someone go in with you because none of your belongings are permitted to go with you to the Operating Room except your glasses and glass case so whoever this person is should be able to stay at least until you go to the operating room so they can take your stuff with them. When you ask this person to accompany you, be sure to encourage them from the start to go home to get some rest once you're in your room because they've had a long day too.

Lastly, don't forget to grab your bag or bags. I ended up with two bags plus my purse. I know you're supposed to leave your valuables at home but with technology today our most valuable asset is our phone and I certainly wasn't leaving that at home! Not when I'd promised an entire list of people that I would text them once settled in my room and you will likely do the same because people care and people want to know you're okay. They want to hear from *you* now, not from someone they may not know.

As long as someone is there with you your stuff will be safe with them anyway. I left my overnight bag in the car to be brought up later. I knew my mom would be coming back with my dad and daughter that evening. If the person taking you will be going back to your home, they could bring your other bag from there later. The tote bag I took in didn't have a lot in it. I put my purse inside that bag. I had also packed an extra pair of socks, the clothes I had worn to the hospital that morning, my glass case and anything else I thought I thought I might want or need to have right after surgery. I, unfortunately, didn't realize that I would be encouraged to get out of bed post-surgically as early as I did and I packed my slippers in my overnight bag. My nurse got me out of bed for a walk mid-afternoon and I ended up with those horrible sock-slippers… but that now gives me the authority to tell you how horrible they are! The socks they supply are one-size-fits-all so they are large and they are very loose and keep falling down. As I advised in the last chapter it's best to take your own slippers with a bit of a rubber bottom so you have grip when you're walking on the tile floors.

Admission & Surgery

When you get to the hospital you probably won't wait long because you were, technically, already admitted at the time of your pre-surgical appointment. My surgery was at 9:00 a.m. and I had to arrive at 7:00 and report directly to the Day Surgery nurse's station. They called me back almost immediately to a private room where I was asked to change into a gown and then my bracelets were put on and I was able to hang out on a stretcher with a nice warm blanket covering me. The nurse took my vitals, reviewed all of my records again, started my IV and told me that at 8:30 someone would come get me and take me to the Operating Room. The nurse I had was awesome... she thought I was a bit young to be having a hysterectomy so I told her what I'd been through the previous year. After hearing my story she assured me that I was doing the right thing.

If the person accompanying you is staying for the duration of the surgery she will be given two cards to hang onto. Each has a different number. One number will appear on the board in the OR waiting room so they can know your status each step of the way. The second card will have a special code so that anyone you give it to can call the floor at anytime to find out how you're doing at any point during your stay. You can give this number to your entire family if you want to. If you have children that can't be there they can at least call for updates.

You may think having to be at the hospital two hours early is a bit much, because I did, but the time passed by incredibly fast! Before you know it someone comes knocking and gets you and your stuff all organized. Because I was able, I had to walk to the Operating Room while others, having other surgeries, were wheeled over. Walking to the OR just seemed weird even though it wasn't the first time for me. It was one of the few moments I asked myself what I was doing even though all the pre-surgical nurses two weeks before and my admission nurse that very morning all assured me that I was doing the right thing. Outside the doors to the OR I hugged my mom and both of us had tears. She was shown where she could wait while I was led to a small, empty waiting room inside the Operating Room ward. That's when I felt the most alone and wished my mom could

Pamela Clayfield

have come in there with me just for the time I had to wait. The corridor outside bustled with the increasing activity of nurses and doctors getting ready to operate. The nurse's station was right there as well, everyone working hard to get organized. I looked for the familiar face of my surgeon but never saw him.

If you didn't meet with an anesthesiologist during your pre-surgical appointment you will see one now. He'll review the questionnaire you filled out and handed in at your pre-surgical appointment and ask a few questions. He'll clarify anything that you marked as *yes* and made comments on, check your teeth and then go finish preparing his equipment. The purpose of checking your teeth really quickly is for the purpose of the Endotracheal (ET) Tube that will be put in place after you're asleep to keep you breathing. He just wants to confirm they are your teeth and none are loose. Finally you will be visited by one of the nurses on your team who will ask a few of her own questions before finally leading you to your Operating Suite where the staff will assist you onto the table and then try to keep you talking to keep you distracted and, hopefully, calm. You won't likely remember most of the conversation. I remember saying I was an RN because I was telling everyone that. They asked what I did. Otherwise I don't remember much of the conversation. You'll remember everything they do as they apply the sticky pads that will monitor your heart activity and breathing and the blood pressure cuff. If the volume is turned up on the monitor you'll hear your own heart racing. My anesthetist was great. He had asked, during our short meeting in the waiting room, about my migraines and I told him that I had one that morning. By the time the Senior Gynecology Resident, who was assisting with my surgery, approached with his checklist, my headache was going away... I like whatever I was given very much! Does that come in a pill for treating all the other migraines I get? The final checklist consists of confirmation of allergies, post-operative medications, all home medications and confirmation that you know what you're having done so they know it's the correct surgery that they've prepped that room for.

My surgeon came in after the resident had reviewed everything with me so he asked how I was doing, confirmed what we were going to do, asked if everything had been reviewed and said "let's get started." I don't remember him walking back out the doors to go scrub in that's how fast I was knocked out. In most cases they will tell you they are giving you the medications and to count backwards from ten, or a hundred. I was surprised I wasn't told to count but I think it's silly anyway so I like how it was done.

I won't get into what will happen after you are asleep. Most people don't want to know. I will say that they move very quickly after you're out. The anesthesiologist will intubate you and hook you up to the respirator. The nurses will insert a catheter, put drapes in place, shave the surgical site and splash disinfectant everywhere in just a few minutes.

Recovery Room

Eventually you'll wake up. You may wake up in the OR as they are transferring you to your bed but chances are you won't wake up until you've spent some time in the recovery room. Your surgery is over. Someone will either be with you or hovering somewhere very close by. Whoever accompanied you is in the waiting room and will only see your status change on the board to Recovery when you are transferred. The codes change according to surgery start, surgery end, recovery room. Apparently it's not the best system in the world. My mom felt a little lost not having the surgeon or someone come out to her quickly just to say that everything went well and approximately how long I'd be in recovery before they transfer me up to the floor and they would call her at that time.

You'll be asked how you're feeling as far as pain goes as well as nausea. They'll check your dressing and even your toes to make sure you have feeling and circulation in your feet. If you feel like talking, which you likely won't, there's not much point in asking how things went. Your current status is their only concern—your surgeon will now be operating on someone else and the recovery room staff won't know all the details except if there were any complications they need to know about.

You'll be in recovery for about an hour if all goes well. Any kind of breathing or cardiac issues or issues waking up will make your stay longer as will excessive bleeding.

Of course there is always an exception to every rule, as in my situation, where the hospital administration had some communication issues.

I kept being asked how I was and if I was okay. Each time I struggled answering because my mouth was so dry. I was finally offered some ice chips. Not only did that help with the dryness but it helped wake me up and I noticed more patients being brought in and thought that I should be going to my room by now. Then I overheard a nurse somewhere tell another that the surgical floor wasn't communicating with empty bed numbers for them to take patients to and they were starting to get backlogged.

Hysterectomies & You

This information, as you can imagine, is not relayed to the waiting room and nobody will go out to let them know... the minutes just tick by and they start to wonder if something is wrong.

Finally someone will come to let you know they have a room for you and it's time to go. They will do a couple of last checks to be sure you are leaving their care in stable condition and there's no bleeding at the surgical site. As they wheel you past the waiting room door they'll call out to your party to join you.

Once in your room the floor nurse assigned to you will be in to see you as soon as she's received report of your condition in recovery. They'll take your vitals (again), check your dressing, IV and catheter and ask how your pain is. They'll also bring you something to drink and I recommend you do *not* drink orange juice... that may have been what brought on my nausea and vomiting. You will be given water as well. Make sure they bring flat gingerale or apple juice, otherwise, stick with the water!

You'll feel really sleepy and this is a good time for the person who came with you to go home and get some rest and allow you to get some rest.

Your nurse will be in to check your vitals every hour for a while as well as your dressing and make sure your pain is controlled. In the time between these checks make sure to rest, if not sleep. It will be incredibly difficult because everyone you know will be texting you or contacting you on social media and you might just have to power down your phone for a while. That doesn't include the unit and the noise from others in your room as well as hallway noise. It was nearly impossible to actually sleep. I was so tired at one point I was in tears and my nurse managed to find a pair of ear plugs for me!

Another issue that nobody will tell you about, and could be a very real problem, is gas. The anesthetic slows down peristalsis which is the natural movement of the intestines that moves food and fluids through your digestive tract. Then your surgeons also move all your insides around which causes the peristalsis to stop altogether. Lastly, you also haven't had anything to eat since at least midnight so there's not a lot to move around in there except air. This stalled gas can cause a lot of pain, sometimes worse than the surgical pain. Your nurse will encourage you to put heat on your abdomen as well as get up and walk. You'll need a nurse to go with you for your first walk, possibly that first day if you're unsteady on your feet. I think more of my pain was from gas than the surgery but I'll never really know.

Eventually your supper will arrive and will consist of fluids including some Jell-O. The only thing not clear is the ice cream for dessert. If you can stomach it, you'll want to consume what they bring partly to help start the process of resetting your digestive system and to start increasing your fluids. The more you drink the sooner your IV can be discontinued. When you are drinking enough they won't remove the IV completely, they'll discontinue the line and turn it into a saline lock just in case it's needed again for some reason.

Don't get too excited… you'll very likely be stuck with the IV overnight the first night as well as the catheter and then it will be dealt with the next morning, sometime after breakfast. Because I was so sick I was grateful because they were able to give me some of my medications, including Gravol, through my IV which was a huge help.

Hysterectomies & You

Finally you'll find yourself feeling worn out. Try to turn in early for the night because your night will come with all kinds of interruptions by your nurse to give you medications and do assessments. Let your nurse know you want to turn out the light and try to get to sleep so they can bring any bedtime medications. I should also mention that you will very likely get an injection of heparin each night you're in hospital. This is a blood thinner to prevent blood clots from forming in your legs due to the surgery and the inactivity of being in bed. These clots can break off and travel to the lungs causing a pulmonary embolism which can be deadly so prevention is key. These shots are given in your stomach and they sting like crazy! Plus they leave bruises at the injection site that stay for so long you'll start to wonder if they're ever going to go away. But, yes, eventually they do go away.

Try to sleep well!

Chapter 8... Time in Hospital

If you had a vaginal or laparoscopic hysterectomy you'll very likely be discharged after one night, maybe two depending on where you are. For an abdominal it will require a second night, possibly a third. Other than my being as sick as I was on the day *of* my surgery, it still baffled me as to why I *actually* had to be there for the second night.

You can almost count on being woken *way* earlier than you would normally wake at home in addition to multiple times through the night. Don't be surprised if you're woken before 7:00 a.m. this first morning because your night nurse has the order to remove your catheter before she goes home. The time you are woken will also depend on when your surgeon is going to pay you a visit. If he has office hours or another surgery at 9 he needs to be making his rounds well before that time.

When your surgeon visits you'll be told how the surgery went and if there were any issues. I actually had my first visit at 7:20... yes, 7:20 a.m.! It was the Senior

Hysterectomies & You

Gynecology Resident who had assisted with my surgery. He informed me that the surgery had gone really well but they had found a significant amount of endometriosis and they'd had to take my left ovary as it was adhered (stuck) to my uterus. I really didn't know how to feel about that because I was supposed to keep my ovaries. His explanation did make me think about some of the issues I'd had for years including my back pain and irritable bowel so I was also grateful though found myself more emotional than I thought I'd be. He also, with my assistance, peeled back my bandages and had a first look at my incision. Of course I looked too. He said it looked great. I raised an eyebrow! All in the eye of the beholder I guess. It was just after I finished my breakfast that my surgeon came in and asked if I'd had an earlier visit. Because the Resident had checked my incision my surgeon didn't feel the need to so we chatted for a few minutes about the endometriosis and the removal of the ovary. He explained that, based on my age, the endometriosis tissue should not grow back. Lastly he said to let either hospital staff or his office know if I was having any issues or symptoms related to menopause and/or feeling emotional. Before he left he also told me he'd see me the following morning for discharge. As he walked out of the room he told me he had been concerned about how sick I'd been and that he had written me an order for a sleeping pill for that night. "Thank you!"

If you've had a vaginal or laparoscopic hysterectomy you'll likely be asked how you're feeling. If all went well overnight you'll be told you can go home and your surgeon will prepare the discharge order. If you had a day or night like I had you will very likely be told you will have to stay another night.

Pamela Clayfield

Before breakfast arrives, your day nurse will come in with fresh towels and washcloths and bring some warm water in a basin so you can wash up. You can wash as much or as little as you want. Just washing your face is a great feeling, especially if you *did* bring your personal face care items along and don't have to use the horrible soap they offer. They'll be quick to clear the basin away again to clear your table off. Breakfast will arrive and it will be normal food as opposed to clear fluids… welcome to the road to recovery and this is your first full day!

At some point after breakfast your nurse will come in and talk to you about a number of things. If you haven't had your catheter removed she'll remove it. If you've had your catheter removed the one thing you'll be told about is what to do when you have to pee. She'll explain that they need to monitor your output to make sure that 1. You're not retaining fluids in your bladder because this is a possibility related to both the surgery itself and the catheter, and 2. Your intake fluid levels are sufficient. She'll show you where your cubby is in the bathroom and how to use the "hat" that's there. She'll ask you to keep track of the amount you go each time. You should have a white board somewhere around your bed so you just write it on there and your nurse will see it each time she comes in.

She will also give you a squeeze bottle to use. They want you to fill it with warm water each time you go to the bathroom then spray your perineum after you use the toilet to clean the area. This will prevent bacteria from moving into the vagina and/or urethra and cause infection. With the catheter just removed the urethra is open because the muscles haven't contracted back to their previous state yet which allows microscopic bugs to get inside and become trapped. What they don't tell you about is the sensations you will feel when you do have to empty your bladder the first time after that catheter is out! Yikes! For starters, it is nearly impossible to go! You can't tell when your bladder is full and needs emptying either. When you sit to go there are control issues at first and you will find yourself merely dribbling. It will almost feel like there's no connection between your brain and the muscles that control how you empty your bladder.

Here's a secret! That bottle they gave you… fill it with nice warm water and spray it over your perineum and it will help you go… the warmth of the water triggers the muscles to start working again. You can expect to do this every time you go for quite some time. If you get the right angle and squeeze it hard enough, it will *not* be caught in the "hat" so won't interfere with tracking your output.

Another thing you will need to track, and be asked about repeatedly, in the midst of having difficulties emptying your bladder, is the amount of vaginal bleeding you have. Also make sure you let them know about having a bowel movement or when you start passing gas as previously discussed. If you're having issues passing gas you won't likely have a bowel movement. But that also means you need to get heat on your abdomen as frequently as possible and be up walking around.

Once your nurse determines that you're drinking enough to shut off your IV, she will turn it into a saline lock you will be able to walk around freely without having to push a pole. As soon as mine was out, I changed into *my* nightgown and started walking around the ward as often as I could. I had already texted my mom and asked her to bring me a real cup of tea that morning because they only serve

Pamela Clayfield

coffee with breakfast, which I don't drink, they don't serve tea and I didn't need a migraine from caffeine withdrawal after everything I had already been through the day before. It used to be better when you got to choose and order your own food. That cup of tea that morning helped to completely get rid of my tummy ache. I felt so much better.

Another thing you need to do, and should be taught to do, is deep breathing and coughing. It's important to get air flowing back to the base of your lungs so that any mucous that might have settled there isn't getting stuck which can lead to pneumonia. Taking deep breaths helps to clear the airways. Take one deep breath, let it out, repeat a second time. After the third deep breath instead of just exhaling force yourself to cough. Getting up and walking decreases the necessity of having to do this because you're opening up the chest cavity by being upright and moving.

Finally, at some point, your dressing will finally be changed. You will finally be able to get a good look at what your incision looks like, if you choose to. The bruising is going to be every colour of the rainbow including black in some areas. You will also likely find bruising on your hips, if you had an abdominal. Every surgeon closes differently so I can't advise if you will have staples, stitches or steri-strips like I had. A much lighter bandage will be applied which will make you feel better too. The initial pressure dressing is heavy, awkward and uncomfortable. The lighter bandage will make you feel like you're not weighed down and walking will be less awkward too.

Moving will be painful and you should adjust your movements accordingly. Getting into bed, and back up as well as rolling over are all painful. You should roll onto your side and use your arms as much as possible. Hospital beds are great because you can put the head of the bed up which is really helpful but at home your bed will be flat so getting used to that is important. Try it in hospital where staff can give you some advice.

As the day progresses you'll find yourself napping, or at least resting, but you'll also want something to do. You'll probably, and hopefully, have some visitors but you'll want to have your book or magazines to read or some puzzles to do. Maybe you're big into video games so you could take a handheld unit or play games on your phone. Just keep the volume down or turned off, or use headphones out of respect for others around you. I found that having multiple visitors tired me out as the conversation moved too fast and I couldn't keep up.

I got so bored. I so badly wanted to go home. Partly because of that boredom, I spent a great deal of time walking around the unit. People started to smile or wave because I passed so many times. I made sure I rested too but it was a long, boring day. I also had a lot of gas pain because it wasn't passing and I was constantly going to the blanket warmer and putting heated blankets on my stomach. Your nurse can also give you something to help move the gas along and don't be afraid to ask for this. It wasn't something even addressed the first day until after supper sometime. The pill she gave me I actually swallowed whole. It worked and that helped make me feel better too… it took the pressure off my surgical site from the inside. The next day she told me to suck on it like a candy. I found it interesting that it didn't have the same effect.

Your pain should be better controlled with oral medications, you should be feeling better, especially if you ended up with post-anesthetic nausea and vomiting, and by the time you head to bed you should have started passing gas.

All of these measures are important if all you want to do is go home.

Have a great, and better, night.

Chapter 9... Going Home!

Your second night should have been much better than the first was especially if you had a good day. Anything is better than that first night was! Just sleeping without the IV and catheter makes it better. You should definitely be feeling better after a better night's sleep with fewer interruptions, and hopefully you started to pass gas sometime before you went to bed or you won't be allowed to go home.

If everything has gone well, you know you should be discharged this morning. I know some hospitals still keep abdominal hysterectomy patients for three nights. If that's where you are, I'm sorry and I hope you have lots to do.

Your hospital stay can be lengthened by complications such as DVT or clots, bleeding, urinary retention... you can't empty your bladder... even though they removed the catheter less than 24 hours after surgery.

You'll be brought towels again but will probably be on your own to get your own water this time because there's probably someone else who is now only one day post-op

Pamela Clayfield

who needs help more. I was so eager to go home that I got my basin and washed up. I even applied some make-up and pulled my disgusting hair into a bun. I had my breakfast and gathered up what I could and packed my bags. I needed to wait for my dressing to be changed before I could get dressed but just set my clothes aside.

Unfortunately you're stuck waiting for the surgeon to find time to get to the floor to get a report on how you have been doing and check in with you. That includes, once again, checking your incision and asking pointed questions about how *you* feel and how you've been doing including if you have started passing gas. If it's the weekend, like it was for me, I had to wait for the on-call gynecologist to come up from the baby floor to see me. You will be asked about whether your pain has been controlled so they know what to send you home with. I noticed that my prescription was actually written by the senior gynecology resident the previous day. Unfortunately your dressing won't get changed until after the surgeon has seen the incision.

Once you have received the okay and the surgeon has written the discharge order and notified staff, your nurse will have to find the time to come in to change the dressing and will tell you to get dressed. While you get dressed and pack your remaining belongings she'll gather up your paperwork and bring it in to you. She will review all of the discharge instructions and give you your prescription to take to the pharmacy… yes, that's correct! Wouldn't it be wonderful if there was some other way? Well, there is, but they don't have time to fax everyone's prescriptions. Whoever drives you home can take it in for you and pick it up when it's ready if you'd prefer. You *could* wait for it if you feel up to it or you could even have it delivered.

Hysterectomies & You

Discharge Instructions

The instructions are straightforward and repeat some of what you discussed with your surgeon the last time you were in his office before you booked the surgical date. The nurse will go over everything in the instructions to clarify, answer any questions and know that you understand. The instructions talk about signs and symptoms of infection which include redness, swelling and heat at your incision; vaginal discharge that is abnormal in any way, like an increase in amount, bright red, green or light brown in colour, passing clots, discharge has a bad odour, and, of course, fever. Recommendations to call your doctor include infection, any issues with emptying your bladder, calf pain that could indicate a DVT, breathing difficulties/shortness of breath and worsening pain or lack of pain control.

A big part that's covered is the importance of taking your pain medication every 4-6 hours to keep your pain under control. Pain management is important because it allows movement and you rest better when you're not in pain. Movement and rest help your body heal.

However, they don't tell you that all narcotic medications cause constipation. It is best to take your pain medications according to your pain *NOT* the instructions on the bottle.

The instructions will also list what you can and cannot do during your recovery. You won't be allowed to drive, do housework, return to work or engage in sexual activities. You will want to spend time resting not only because you will need the rest but also because doing too much will also cause an increase in swelling at the surgical site which, in turn, causes an increase in discomfort. You will also not be allowed to lift anything heavy so you can't even do laundry, vacuum or change a bed. These activities can resume after you get the okay from your surgeon.

Looking after yourself and your incision and surgical site is key to recovery. That peri-bottle you received will come in handy and you are to continue using it at home as long as you are having vaginal bleeding/discharge. This will likely be more and for longer if you had a vaginal hysterectomy. Bleeding after an abdominal hysterectomy is less. Continue wearing pads as tampons are not allowed either. Why? You ask... because you don't want that discharge sitting and soaking on a tampon right at a surgical incision... don't forget you will have an incision that has closed off the top of your vagina. This will take longer to heal because it is a warm, moist environment.

You should be given multiple forms of pain medication. A narcotic such as Percocet/oxycocet or Dilaudid/morphine as well as a high dose anti-inflammatory like Naproxen and acetaminophen. Unfortunately, all narcotics tend to cause constipation which is something you do *not* want to experience post-hysterectomy! Constipation will be discussed numerous times. The best option is to start taking something to prevent constipation from starting right away. Your surgeon may prescribe something like Senokot which is a mild laxative. Alternatively you can pick up some docusate sodium (Colace is one of the brand names) which is a stool softener. You can take one a day up to two capsules twice a day if needed. If you continue to have issues passing gas then try peppermint tea or Oval. Your pain will decrease in the first few days and you shouldn't need them very long or as frequently as ordered.

You will sign that you read and understand the instructions and that you understand you need to have a follow-up appointment with your surgeon usually around the 6-week post-op mark. This may be longer depending on the surgeon's schedule in the office.

You're home!

You'll get a wheelchair ride to the exit and you will need to take it easy getting into the car.

Travelling home is not fun, unless you're lucky enough to live right next door to the hospital. Even the smallest bumps in the road are highly noticeable when you have had surgery and any movement of the tissue causes pain. For the majority of my trip home I said "Ouch." Try to brace yourself in the car so that your arms can take some of the movement when the car goes over a bump.

Just like when you've gone on vacation and arrive back home it feels great to walk back through your front door and be in your own home. We all like to be home! If you have stairs that you're concerned about don't be. They're not actually too bad. I did think that I might come home and end up spending most of the time in my room but it wasn't necessary.

When you get home, you'll probably be exhausted and want to go rest. It's a good idea to rest when you feel the need to do so. Of course you'll also want to unpack your bags and settle back in but you may just have to rest first.

You also haven't showered in 48 hours or more and, yes, you are allowed to shower. I was incredibly eager to get in the shower and clean up head to toe because washing from a basin just isn't the same and my hair was totally gross. It's recommended that you have someone present the first few times you shower just in case you need help. You are not allowed to have a tub bath, use a hot tub or go swimming. You can't soak the incision or it could split open because soaking in water softens the tissue. Pat the incision dry, do *not* rub it!

My dressing in hospital was actually changed to the exact same Tegaderm pad that I previously mentioned I had purchased prior to my surgery. They recommend changing your dressing every 24 hours and you are able to leave your incision uncovered if it's not irritated by clothing. This means that you could shower the day you come home from hospital and then shower the next morning before you take the dressing off to check your incision. These bandages are quick drying so they are great for this use. You could then leave it open until the following morning as long as it's not bothering you and cover it before your shower again. Leave it covered and take it off after your shower the next morning. By that time you have showered four times and can evaluate whether you feel it needs to be covered after that point for showers or if you're comfortable leaving it uncovered and taking extra care with the water spray.

Of course you won't have anyone bringing you a meal tray three times a day so you'll have to figure out what it is you're going to do about meals. Hopefully you took my earlier advice and prepared some meals ahead. You also won't feel very hungry so you'll find yourself forcing yourself to eat three meals a day or more, smaller meals. Make light meals like soup and try to pack in fiber and protein. This is why I talked about making meals ahead of time because you could make something with chicken or beef like stew. Don't be afraid to eat half a sandwich or have toast with peanut butter and jam. Try to keep up with the four food groups and keep up the protein to assist in healing.

Contrary to popular belief, healing uses lots of calories and energy. That's why eating important as is eating food high in protein. Vitamin C also helps promote healing so orange juice and vitamin supplements are helpful. Pick up Vitamin C tablets at 1000 mg and take at least one a day especially if you're drinking juice. Two tablets a day if you're staying away from the citrus.

Also make sure you keep drinking water which will help flush anesthetics and all those medications out of your system, as well as aid in healing and help with the constipation.

Just Be...

You'll be really glad to be home. Settle into your own routine; a new routine. Do the little things you rarely have time to do like read, do some puzzles, knit or crochet, watch TV and movies and revel, for at least the first little while, in the peace and quiet of home compared to the hustle and bustle of the hospital.

Chapter 10... The Initial Recovery

Period... 6-8 weeks

Some of you may have picked this book up just to get to this chapter. You want to know what to expect post-surgically and I get that, totally. Post-surgical was what I worried about the most. When you signed up for this surgery nobody told you what you were really in for and exactly what you were going to face. The main reason I decided to write this book is to bring awareness of what to expect so hopefully all of you can avoid all the surprises that I experienced and am still experiencing. I figured if I was surprised by so much and I had an idea of what to expect then there were more of us that needed reassurance and education. For the remainder of the book I have done my best to cover as much as possible in order to answer as many questions as I can.

Hysterectomies & You

In the days following your surgery you will experience a fair amount of discomfort in a variety of areas. Of course, as expected, the surgical site will be the most tender and painful. It will be bruised many different colours as will the bruises from the blood thinner injections. The bruising could reach all the way to the front of both of your hips and the tops of your legs. There will also be discomfort over the front hip bones to the extent you will not be able to touch them.

My pain, overall, wasn't too bad. I've spent a good part of my life experiencing back pain so maybe my threshold is higher. Pain is also very individual what I feel is excruciating may be a dull ache to you and vice versa.

It's important to take your pain medications but it's also important to try not use too much of the narcotics because you don't want to take medication that you don't need and you don't want to run the risk of addiction. As discussed last chapter, your surgeon will also give you other pain medications that don't contain narcotics and it's wise to make use of these. IF they provide these medications on your prescription it won't be in the doses you were getting while in hospital so you may want to check with your surgeon or family doctor to confirm that you can continue taking Tylenol Extra Strength 1000mg and ibuprofen (Motrin or Advil) 800 mg at the same time. When taking these doses it's safest to keep it to three times a day. I was discharged with a prescription for Naproxen 500mg to take twice a day so I took two Tylenol Extra Strength three times a day. Acetaminophen and anti-inflammatories increase the effects of each other. There are plenty of options for pain control and out of 30 morphine tablets prescribed I used 17. I was taking a single morphine in the morning when I was

Pamela Clayfield

showering, making my bed and generally more active. Then I was taking another one at bedtime to help me get to sleep. I stopped the morning dose but continued the bedtime dose for a few extra nights. Otherwise I was using the Tylenol/anti-inflammatory combination very effectively. I was also applying heat over my surgical site a couple times a day.

Constipation is a huge issue and can actually cause more pain than the surgery itself. It doesn't matter which surgery you had done because the same cuts have been made inside and your colon passes the same surgical site. Whether you have a history of constipation or not doesn't matter. Narcotics will cause it. It's best, as I keep mentioning, to be consistent with taking a stool softener or mild laxative.

You will find that tying shoes or boots is nearly impossible so I recommend getting pull-ons. I wished I'd had my surgery in the summer when I could have just pushed my feet into flip flops. As for clothing, wear whatever you're comfortable in and you should try to get dressed every day. It's easy to fall into a pattern of just keeping the pajamas on and then we develop a mental slide in that direction as well. Zippers will probably be a bother as will anything tight at the waist due to the swelling. Leggings were easiest for me and only for a short time each day to start. I was able to wear them for about five hours and then I was back in my nightgown.

You'll spend a lot of time just resting… on the couch, in bed… wherever you can get comfortable. Support yourself with pillows. You'll find it difficult to sit upright for long periods. They've cut muscles so core strength is

Hysterectomies & You

diminished. For that same reason rolling over in bed is almost excruciating. One of the best excuses to rest!

Make sure you get out and do some walking. I was forcing myself to walk around my complex just to get out, get some fresh air and move. Movement helps keep the bowels moving, including passing that gas which, even weeks after surgery, can still build up and be very painful against the surgical site. Of course this depends on the time of the year. My surgery was in November so I was lucky that there was some mild weather... for about a week. After that I was stuck indoors.

You'll likely have visitors... try to arrange for times when you're not so fatigued... you'll learn this in the first few weeks home. For me, by 4:30 I was exhausted.

You will watch your incision slowly heal. At first, aside from the bruising, you will notice that your incision looks very angry. One of the first things you will want to get rid of is the steri-strips. These will slowly curl and you can trim the edges making then shorter before they finally fall off completely. If, at some point, you notice that a strip is only sticking to one side of the incision, you can go ahead and peel it off, *very* carefully because *all* the tissue is very tender. If it's not sticking to both sides of the incision it's not doing its job anymore so it may as well not be there. You're not supposed to put anything on the incision and I didn't for about the first five weeks. After that I started putting just a small amount of antibiotic cream on it. The ends of the incision may not have closed or been closed properly and these may appear to be open and oozing. If there's a suspicion of this, cover that small area with a Band-Aid and antibiotic cream. Keep an eye on it. This was something I ran into a couple of times and it worked beautifully. I didn't want to have to be seen and treated for infection.

Other infections you need to watch for are both bladder/urinary tract and vaginal. It's very easy to get a bladder infection after you've had a catheter. Despite all the measures I advised of earlier to get your bladder emptying after the catheter has come out including keeping the area clean, sometimes it's just difficult to empty completely and there's less feeling than usual. There may have been just a small amount of bacteria introduced at the time of the catheterization or shortly after it was removed before the muscles at the openings could start to fully function and seal properly again. Monitor yourself for fever which commonly shows up as feeling cold followed by feeling hot or chills alternating with sweats. Of course, if you lost your ovaries, hot flashes could be mistaken for a fever so grab a thermometer if you're not sure. Other symptoms of bladder infection can be the feeling of having to go (urgency) very often (frequency) but very little comes out, burning or pain when you go pee, or actually seeing blood in your urine. Everyone's bladder infections are different so it's very hard to say how it might be for you. Anything unusual should be concerning.

I started with a low-grade fever and thought I was getting a cold. By late in the evening I was actually leaking urine. I had no control. I'd never had a urinary tract infection before so I had no idea how it would manifest. I was terrified that a small hole, called a fistula, had formed between the urethra and my vagina because that's where it seemed the urine was coming from. It was the weirdest thing and I called the surgeon's office, but he was closed, followed by my family doctor who asked me to get to a lab to give a urine sample and they sent a prescription to the pharmacy for me with the assumption it was a bladder infection. They did confirm that's what it was and after a few days on antibiotics the leaking had stopped… what a relief.

Just as I was finishing those antibiotics I started noticing extremely foul smelling vaginal discharge and I called my surgeon's office. This time they were open and I explained that I was just getting over a bladder infection but was experiencing a tan coloured, extremely foul-smelling vaginal discharge now. He started me on a second antibiotic. After the two courses of antibiotics the infections were gone and I started to feel much better, much quicker. The problem with having infections though was it slowed down the healing process at the surgical site and probably set me back by two weeks.

Another thing to continue monitoring is vaginal bleeding and/or discharge. The discharge should never be excessive or heavier than a period. If you start bleeding and have to change your pad every 45 minutes or less you need to go to the emergency department. Be aware of the discharge you are having in hospital. Once you're home it should slowly decrease. If it changes colour, such as the tan colour I had, or green, or it starts to turn bright red again or you pass clots you need to call your surgeon. I was five weeks post-op when I started bleeding. It was definitely fresh blood and I was worried, four days before Christmas, that it was another infection forming. It did stop on its own before I heard back from my doctor which was a relief but, again, it's still worrisome. Your discharge could also get thin and start to smell bad and that too needs to be called in. When it comes to discharge there will be a time when it does

Pamela Clayfield

change colour but it won't have any smell. This could be granulation tissue. Granulation tissue forms to allow healthy cells to form underneath. You have an incision in there that needs to heal. The granulation cells have to, eventually let go, and come out in your discharge. Eventually the stitches they put in there will start to dissolve and these, too, need to pass. You may notice pieces of thread and you may not see a single one.

Fatigue seemed to be my biggest hurdle as well as tiring easily and quickly. I was okay in one-on-one situations but if there was even a small group where there was a lot of noise and conversation I found it tired me out very quickly. I couldn't focus on multiple conversations. This is why my advice earlier in the book was to think about what might be happening in the weeks following your surgery because it's possible you may experience something similar and lack the focus to attend certain events.

Whether you kept both ovaries, lost one or both you will be told to monitor how you are feeling emotionally. It's sometimes difficult to discern emotions especially if you had your hysterectomy because of a uterine cancer and not only lost everything but have to go through menopause because you can't be prescribed estrogen and you have to face treatments. I was starting to have night sweats and my emotions had felt like the switch had been turned off. It was the best and only way to describe it. I have always been a sexual person who gets butterflies when two characters on TV kiss for the first time after all that tension build-up and I was experiencing nothing. I also felt nothing for my significant other which left me feeling guilty and wondering what on earth was going on. I couldn't wait to see my surgeon.

Chapter 11... The Recovery Period is "over"

No. It's. Not.

Six weeks goes by quickly and you do feel stronger every day. You will watch your incision become less angry and turn into a red line or your small laparoscopy incisions come close to disappearing. Your pain will diminish and you'll probably be bored, among other things.

Six or eight weeks is just NOT enough time and you may need to evaluate how *you* are feeling before you decide to return to normal activities.

Everyone else will quickly lose track of how many weeks post-op you are. They will think your surgery was long ago and they'll tell you how great you look. I don't know how many people said "You look great! You had surgery when?" They don't realize you still have significant

Pamela Clayfield

healing going on inside, because like you, they can't see it... but *they* can't feel it either. You'll know exactly how many weeks ago your surgery was *and* you'll be reducing that time frame if, like me, you spent any time on antibiotics. When I went back to work six weeks after my surgery I was, in healing time, probably four weeks post-op corrected. It wasn't good and I should have spoken up and said how bad I felt.

You will have started to do things again and even push it a little bit... you're probably doing some laundry but carrying it in your arms rather than using the basket or hamper. Maybe you've cheated and driven to the store. Once you've seen your surgeon it won't be cheating anymore!

You likely won't be napping anymore but you'll still tire quickly and easily and bedtime can't come soon enough, especially if you had to go back to work.

In a perfect world you would see your surgeon at the 6-week point and get the go-ahead to return to normal function. However, with surgical schedules always changing as well as your own schedule to work around this will likely not be the case.

When you do see your surgeon he will, naturally, ask how you're feeling. Make sure you take a list of questions and concerns to this appointment and that list should include anything you have found to be out of the ordinary or that goes against what the hospital staff told you at discharge or your surgeon when you saw him last in hospital. He will check your incision(s) and he will have to use a speculum to look inside the vagina to check the incision in the vault as well. He'll also review the Pathology Report and let you know if there was anything abnormal found. Pathology reports take about two weeks. If you don't hear from your surgeon's office *before* your 6-week follow-up then it's safe to say that the pathology report didn't show anything significant. I was happy to know that there were no abnormal cells. There was no sign of them anywhere.

Unless he finds something on exam, he will likely tell you that it's safe to resume all activities, including sexual activity, but won't elaborate any further on what you may experience... another reason I wrote this book.

If there is any granulation tissue in the vaginal vault you may be advised to hold off on sexual activity for another few weeks because it obviously hasn't finished healing. If healing isn't complete you will very likely know because you should still be using a pad or panty liner for mild, ongoing discharge. There are multiple factors that slow the healing process. Age, not following the discharge instructions, health conditions like diabetes that slow healing and simply just the location of that incision... a warm, moist environment that doesn't promote quick healing can play a huge role. It's different than an external incision that can be exposed to air and is able to dry out.

Be sure to address any emotional and physical changes you've experienced. By the time you get to this appointment hopefully you have overcome any feelings of loss. If not, you do need to bring this up so it can be managed before it leads to depression.

I asked how long it usually takes for a single ovary to take over hormone production for both ovaries. His answer concerned me. "Right away, why?"

I explained my night sweats and didn't even make it as far as telling him about my emotional state... how I had been feeling, or rather *not* feeling. I told him I'd had to

Pamela Clayfield

change my flannel sheets back to cotton and turn down the heat. "If you ask anyone they'll tell you I'm *always* cold!"

He was concerned, in fact his non-verbal cues indicated he was more concerned than he was letting on. He told me the ovary looked to be in good condition when he'd left it in but perhaps it wasn't up to the task. He gave me a requisition to go for blood work to check a number of my hormone levels. If something similar is going on your doctor should check your Follicle Stimulating Hormone (FSH) and Estradiol (Estrogen) levels to see if you're post-menopausal or not.

Sure enough my remaining ovary had, in fact, failed and I had also gone through menopause in 6 weeks or less, a process that as some of you know, normally takes years. As soon as he saw the results he didn't hesitate to start me on Estrogen. He had the prescription to the pharmacy within hours.

A note about estrogen… don't let the pharmacist tell you to avoid taking it with calcium. I didn't take it the first night because when I went to pick it up the pharmacist told me to take it with supper because morning is when we usually have calcium (milk). I'd *never* heard that before. I checked with a friend who had been on it for years and I checked with a doctor friend and nobody could understand why I was told that.

I did start it at bedtime the next night and had no issues even though I took it after I had eaten a bowl of cereal. When I was back in the store I spoke to the pharmacist and store owner that I know very well and usually deal with and she'd never heard that estrogen couldn't be taken with calcium. She and a pharmacy technician searched their databases and found nothing. She didn't know why I had been told that because the pharmacist I described to her was usually really good with that type of information. So I feel it's important that you are aware of this as well. If the same thing happens to you, I don't want you to hesitate.

Not everyone is able to ask the right questions and everyone will experience different post-op issues. You may experience issues that I haven't included here because I didn't know about them. I did my best to talk to other hysterectomy patients about their experiences so I could provide as much pertinent information to you as possible. Excess pain from constipation, ongoing tenderness and swelling and ongoing fatigue seem to be the biggest and most common complaints.

I started writing this when I was almost four months post-op and, after editing, I finished around eight months. I started to write this when I realized I was one of the lucky ones. My education helped me to understand the surgery, the changes and the symptoms as well as what to expect during the recovery period. It also helped me to ask the right questions based on what I was experiencing. Working in family practice has taught me that not everyone has that knowledge or experience. I want to cover off a number of issues that you will likely experience to some extent even after your 6-8 week post-op appointment when you've been given the all-clear.

One of the issues I'm still experiencing at eight months post-op is varying sensations at the surgical site around my incision… tingling, numbness, burning and sometimes stabbing pains and at times the burning and stabbing pains feel very deep. Other times they feel very superficial… at the surface… sometimes it feels itchy just under the first layer of tissue and all the scratching in the

Pamela Clayfield

world doesn't relieve it. I also still have tenderness at the site and can't lean against a counter; but yet I can sleep on my stomach. The pain at the front of my hip bones varies and started to improve around five months post-op.

You will have ongoing swelling. I asked my surgeon about this and he said it's normal and can last for a year as can the tenderness and discomfort at the surgical site. It's commonly related to activity and caused by the contents of your abdomen being handled and moved around during surgery as well as the trauma of the incision through multiple layers.

Fatigue is also an ongoing issue. It gets better but even at eight months post-op I still feel tired at the end of each work day. I still rest as often as I can and haven't completely returned to my usual visiting, errands and activity levels in the evenings and on days off. I *have* returned to writing more which allows my mind to work and my body to rest.

I will also warn you about the impact your bladder and bowels will have, again on an ongoing basis. If you have the urge to empty your bladder and you try to hold it, the pressure builds up fast. Unfortunately this pressure builds up just behind the surgical site and tends to cause discomfort. The urge to empty your bowels will be very similar and, in some ways, I find it worse. Constipation has been worsening for me and so has my discomfort. If you have a history of constipation I encourage you to continue taking a stool softener every day because the colon is on three sides of the surgical site. You may find alternating between a stool softener and Senokot might be more advantageous but what works for one person doesn't always

Hysterectomies & You

work for another so I encourage you to try various combinations.

For the remainder of the book I will discuss a number of potential questions that you may have. I have also saved an entire chapter for emotions and sexuality because of the impact this surgery has on both of those areas and the likelihood that you want to know but just may not be able to bring yourselves to ask because it's personal and many people find those topics embarrassing or difficult to discuss.

Chapter 12... Hysterectomy &

Beyond... Q & A

It doesn't matter how old you are, where you are in your life cycle, if you've had no kids or five kids, the reason for your surgery or the length of time between having the discussion with your surgeon about the surgery to the surgery date, you will have questions racing through your mind. There's no way you can ask all of them because you won't know before surgery what you want, or need, to ask.

I have based the following Q & A on the issues I had as well as my questions that and an internet search on *Commonly asked questions about hysterectomy.* I also asked others about their experiences.

Could I avoid hysterectomy?

Sometimes, yes. What is your diagnosis? If you did not initiate the conversation about hysterectomy like I did then your surgeon probably feels it's necessary. What is his rationale? Don't be afraid to ask this question. It's *your* body. A lot of people don't realize that their body belongs to them and you don't *have to* accept what a medical profession advises. There will always be surgeons who prefer to do surgery rather than try to manage symptoms or offer alternate treatment options. Gynecologists make these kinds of decisions all the time often with two lives in their hands not just one.

Heavy periods can be treated by ablation, a surgical procedure where the lining of the uterus is frozen or burned. This heals to scar tissue and prevents the lining from building up during your cycle and, of course, it then doesn't shed, or sheds much less during your period.

Fibroids can be treated by either removing them or cutting off their blood supply... is this a permanent fix or just temporary with the potential need for a hysterectomy in the future anyway? Those are the key questions because most people want as few surgeries as possible in their lifetime. Surgery is invasive, risky and disruptive to life's routines. Would you want to have a second surgery? If the chance of having a hysterectomy anyway at some point is relatively high it's probably worth doing it in the first place.

Be sure, if you really want to try to avoid hysterectomy, to ask your surgeon if there is any alternative, what it is and the outcomes both short term and long term as well as what the chances are of ending up needing a hysterectomy down the road anyway.

How long is the recovery?

As discussed throughout the book, recovery time varies for each person and each type of procedure and your post-surgical instructions will have that clearly outlined. Because of the internal cuts I still recommend planning for 6-8 weeks of decreased activity. I still feel like I'm in recovery mode even at eight months post-op.

These guidelines are just that and it doesn't mean you will be completely back to normal by two months after your surgery. That timeframe refers to the main healing that your body needs to go through at the surgical site in order for you to not do further damage by driving, heavy lifting or generally overdoing it.

Fatigue and swelling can last for a year and possibly longer. When I asked at my post-op check about the swelling I was told it was normal and could last for a year. I'm sure it could last longer because we all heal at different rates and we all have different activity levels. Even at eight months post-op I'm having a significant amount of swelling... it's still *very* noticeable at the end of the day. Even if I'm doing a lot of walking it gets worse.

With the fatigue, it's important to rest when you feel tired which isn't always possible, especially if you've gone back to work. Try to create a workable plan... you may have to cut back on certain activities for a while to get as much rest as possible.

Transitioning back to work is best if you can start off part-time and build yourself back up to your full hours to allow you to rest. However that's not always possible so try not to fill up your schedule around your work hours in the beginning. Getting others to understand is difficult. They

Hysterectomies & You

take your advised recovery time literally and assume that you will be 100% by the date you expect to return. It's better to have a part-time plan in place and get back to full-time sooner than planned than to end up worn out at the end of the first week.

It's also advised to continue not lifting heavy items longer than they suggest. Go ahead and do laundry and vacuum but try to not lift anything over 10 lbs for as long as possible. This will give the tissues longer to heal and get stronger again before they are strained.

If I haven't gone through menopause, will I go into menopause?

It's probably best to start off with a brief lesson about menopause before I answer this question. This excerpt is from an article I wrote just a few years ago…

"Very few women are lucky enough to have their ovaries just turn out the lights one day with no warning or symptoms. Most go through the fluctuation of the hormones progesterone and estrogen, similar to the lights flickering during a storm. All of this generally starts, unnoticeably to most, in the *mid-thirties* when the production of these hormones starts to diminish gradually. This leads to irregular periods, changes to vaginal lubrication, mood swings, lack of energy and interest in sex, and emotional changes like depression and irritability.

Why? There are estrogen and progesterone receptors throughout the body, including the brain, which tries to compensate by producing the mood-regulating chemicals serotonin and endorphins.

Pamela Clayfield

Ultimately, this phase of your life, commonly known as *peri-menopause*, starts around 35 and could last up to twenty years."

Before my surgery and subsequent outcome, based solely on my education, I would have said, yes if they're taking both ovaries; No if they're leaving one or both ovaries. Was I *ever* wrong!

I was left one ovary and, likely due to my age and possibly just the trauma of surgery, it shut down and I went into menopause. There are no set rules. Each of us is different. I'm sure if you're young (under age 35) they will do everything they can to save at least one ovary and it *should* continue to function. Because menopause technically starts around age 35, anytime after that would be questionable, as in my situation.

What are the hormonal changes due to a hysterectomy?

With any surgery, the surrounding structures are examined because if there's something wrong it's best to deal with it at that time rather than having to do a future surgery. Therefore, if you were scheduled for a total hysterectomy (removal of uterus and cervix only) your ovaries and tubes will still be examined. As long as there is nothing abnormal identified then they will be left and should continue to produce hormones the same as before. If your ovaries were removed then you will experience the symptoms of menopause which could include vaginal dryness, hot flashes, night sweats and sleep problems. Because the ovaries do continue to produce some estrogen, progesterone and testosterone after menopause, but in greatly reduced amounts. even if you were post-menopausal before your surgery you will experience the symptoms of these hormones now being completely absent. Whether it was a planned oophorectomy or they visualize a problem and remove both, if you were pre-menopausal you should still be started on estrogen post-op. There can be risks associated with estrogen therapy including blood clots, increase risk of heart disease and stroke. It's important to evaluate what the greater risk is and treat accordingly. In my case, they left me one ovary and assumed it would continue to function normally. My assumption is that the ovary may have been functioning poorly or not have been working at all prior to my surgery.

What about well-women care post-op?

Unless your cervix is left behind, the preventative care programs say pap tests are no longer required post hysterectomy. However, there is always a rare chance of vaginal cancer so regular screening should still be performed up to the age of 69 to monitor for any abnormal cells.

Additionally, after menopause, whether naturally before surgery or surgically induced with no estrogen therapy, bone density screening should be started because estrogen supports bone health and is no longer being produced. Making sure your calcium intake is adequate, at least 1000mg per day and taking Vitamin D, at least 1000mg per day will be a big help as will low impact exercise like walking or cycling.

If you are started on Estrogen, it will help maintain bone health and bone density and is less of an issue.

Pamela Clayfield

How will I know I'm going through, or have gone, through menopause?

As you have probably heard from every woman who has ever gone through menopause including your mother and grandmother, hot flashes seems to be the biggest tell and has become quite the joke over the years. Night sweats usually start and can be quite alarming because it *could* also be fever especially during the initial post-operative period. If you're not expecting it then it's hard to know the difference. The night sweats and having absolutely no desire were my only symptoms.

Other symptoms include vaginal dryness which can also lead to discomfort or pain during sex. You may experience emotional changes that could be more drastic than with the usual hormonal fluctuations during your regular cycle. Lastly, you could have difficulty sleeping as well as dry skin, eyes or mouth which I experience even after being on estrogen for six months.

What does it mean if I have brown coloured discharge post-operatively?

There are many reasons why you may discover, on a trip to the bathroom, brown discharge on your pad or liner. It's a matter of discerning whether it requires a call to your doctor or not. Initially it can be alarming especially if your discharge is minimal or has completely stopped and then there's discharge again.

Blood that is 'old' is brown in colour. Usually this is dark brown, has little odour and dries quickly. Of course

83

Hysterectomies & You

there will be lots of semi-dried blood that has to go somewhere as you heal.

About two weeks after surgery the stitches that were put at the top of the vagina will be dissolving and these, too, need to exit the body. They may have a slight odour because they often do and they're bringing some dead tissue and cells with them.

Granulation tissue sloughing off as the vaginal cuff heals and healthy tissue grows back will also be a cause of brown discharge and this, too, is normal.

It is brown — light brown, dark brown — with a bad odour that you really need to be concerned about. This is like when you smell something and you wondered what died. A fever could also accompany this but if the smell is that bad, do *not* hesitate to call your surgeon or doctor.

How do I know I'm actually ready to return to work?

From everything I have read and everyone I have talked to, 6-8 weeks is the maximum rule of thumb no matter what procedure you have. There are after effects of the anesthetic. Your body has had a lot of trauma done to it and all of that lowers the immune system and healing of other things like bacteria, viruses and even a paper cut or bruises as it concentrates on healing all of the wounds created during surgery.

Take it slow. Talk to your employer if you can. Explain that you'd like to come back but may only last a few hours. Gradually work yourself back up to your regular hours. You will very likely feel exhausted by the end of the day no matter how you progress back to full hours. If you

Pamela Clayfield

have a physical job you may wish to consider more time off so you can hold off on lifting and allow more healing time.

What are options for ongoing pain/discomfort, fatigue or bladder issues if the uterus was removed due to prolapse?

It's difficult to know what was all done inside during the surgery unless you watch it on YouTube. I actually saw a hysterectomy in nursing school... it was one of the two surgeries I got to watch as a student. Everything inside has been shifted and shifts more after surgery and has to resettle. The biggest concerns are whether adhesions form as the affected tissue heals and attaches itself accidentally to surrounding tissues.

Finding a physiotherapist who offers pelvic floor therapy can be helpful. Talk to your doctor or surgeon about this as an option and feel free to ask questions of the physiotherapy clinic or the therapist to see if this could be an option for your situation.

Additionally, naturopaths or osteopaths can be helpful with fatigue and discomfort by making suggestions about what general changes you can make to lifestyle, diet and therapy to reduce symptoms.

There are also plenty of forums online to help answer some of your questions. I find hystersisters.com to be helpful because it's the first-hand experiences of others. It's not professional advice but some of the professional advice sites I have visited have made me raise an eyebrow. They are either too factual and difficult to understand or the information is based on a single person's experience. I have found I took what I read online with a grain of salt and did further research when I doubted something.

Talk to people who have had the surgery if you can. This is a better source of information on healing than doctors and nurses because people can't always share their symptoms and issues with their health care professionals or they have tried to but been unsuccessful as they are told *"it's still too soon after surgery to know. If you have problems in a few months, come back and see me again."*

Pamela Clayfield

Chapter 13... Emotional Changes

Post-hysterectomy

There are many emotions at play when it comes to hysterectomy.

Yes there are the chemically induced emotional changes that women experience with every monthly cycle but it goes deeper than that.

We are women. We were born with a uterus and ovaries which allowed us the ability to carry and bring a child into the world. Many of us have and, that too, was an emotional time.

Now we have reached the point where someone has told us that removing those parts would be of benefit for one reason or another. Maybe it's to save our life; maybe it's to reduce a great deal of pain or relieve anemia due to periods that are heavy almost to the point of hemorrhage.

Hysterectomies & You

Either way we have to come to terms with the loss of these organs and those feelings can be about that loss or they can, for those who are pre-menopausal, be overwhelming as you deal with knowing you can never have another child.

Loss isn't just related to death like most people are led to believe. Loss is related to anything we lose or experience whether it's a marriage that falls apart or an assault or theft.

Everyone is different and everyone handles loss differently and everyone has their own timeline to grieve.

I was not impacted by the feeling of loss related to the surgery. I understood why I was asking my surgeon about having this surgery and I knew I was done having children, a conscious decision I had made years earlier because of my back condition. I want to be around to one day see my grandchildren so it was time to remove the threat. My training also helped me to understand and sometimes facts and understanding are all that stands between working through emotions to acceptance and letting something like this happen and be accepting of it.

I was actually more emotional when I learned I had gone through menopause. It came as a shock, one I hadn't given a moment's thought to prior to surgery. He had no intention of it happening. Clearly I wasn't expecting it and that took some time to grieve, accept it then adapt. At least the estrogen helped relieve some of the worst symptoms like the night sweats and lack of emotions.

If the emotions or depression are bad enough you need to seek counselling or talk to your doctor or someone else you can trust. Grieving any loss is the same and you will experience all the stages: anger, bargaining, depression, acceptance, etc. You will cry. It's all okay.

Pamela Clayfield

Be aware of your emotions because you may feel blue but you could easily slide into depression. The blues should feel similar to post-partum blues and last for a few weeks but if it lasts longer you need to speak to your doctor. This could be hormone related or it could be a depressive disorder that needs to be addressed, possibly with medication to manage it.

A sense of loss regarding sexuality may also create additional feelings and add to your grief. If you have always been a sexual person and felt intimacy was an important part of your relationship the idea of suddenly going into menopause can be terrifying and overwhelming. You may start to mourn the loss of intimacy as well.

I will address these changes in the next chapter on sex and sexuality because there are ways to work around the issues you think you will have and it's important at this time to be connected to the person you love. Your significant other is sharing in this experience with you and may just have similar thoughts and feelings. They, too, may be experiencing grief especially if there are any feelings of being pushed away or no longer being able to engage in intimacy with you.

Chapter 14... Sex after Hysterectomy

I realize that when it comes to sex we are all different. We all come from different backgrounds, were raised with different beliefs, and have varying experiences. I am writing a book that will be read by multiple generations and age groups not all of which are open to discussions about sex and relationships and others who have had little interest in or little desire for sex for a while or who have lost their life partner and are alone with no intentions of another relationship. I also know, and hope those in that category realize that there will also be women in their 30s, 40s (as I am) and 50s reading this book who will want to know if their sex lives will be impacted and, if so, how. Some of you have lost interest in sex for a number of reasons and I hope that your surgery doesn't become yet another reason to no longer have, and enjoy sex.

I recently saw the movie *The Book Club*. I love books plus I'm a writer and love romantic comedies. It looked too good not to see and looked too good to wait until it came out on DVD. It was awesome! It makes you realize that sometimes sex isn't dead, it's just dormant. So this chapter has been written for those who need the reassurance and for those who may just be curious.

Throughout the book I have discussed how hormones play a significant role in our emotions. Our emotions play a significant role in all of our relationships whether we are classified as huggers when it comes to our friends and, of course, being intimate when it comes to our partners. Hormones help to stabilize our emotions and they help our thoughts and feelings of desire.

Watching or reading a love scene has always given me butterflies deep down in response and sometimes been a source of tears for me. That's just who I am.

Whether your ovaries have been removed or one has or the ovaries have decided to stop working, your hormone levels suddenly drop to almost nothing. This can make you feel like the switch has been turned off on *all* your emotions. I truly felt lost. I didn't care whether I got a hug or not... from *anyone*. I felt like I was just kind of existing and my body was healing.

This should eventually sound an alarm especially if you've always been a very sexual person... you enjoyed sex, you found pleasure in sex and intimacy. The switch being turned off and leaving you in the dark is incredibly frustrating because you want to reach out to your partner but your emotions and your body isn't responding and you just want to know what's going on.

In the short month I had to prepare emotionally for my surgery I never once thought about surgical menopause... that wasn't the plan and I had complete faith in my surgeon. I obviously needed to have faith in my own body! I did, in the middle of the night one night, suddenly wake up and wonder if I would ever reach orgasm again. I don't know where that thought came from but it was something I needed to know. The search online only concerned me and I regretted looking. I tried to forget what I had read and that the majority of sites were related to the 'type' of orgasm and mostly talked about uterine or cervical contractions with orgasm and how having a hysterectomy removes these organs and therefore the ability to orgasm. I decided to put that out of my mind and focus on the surgery because it was more important.

Not only do hormones have an impact on our emotions but they play a big role in our sexuality. If you end up in surgical menopause your ovaries greatly decrease the amount of hormones they produce... *all* hormones... Estrogen, progesterone and testosterone... yes, women produce testosterone too and it plays a role in our desires.

Estrogen keeps the vaginal walls healthy, flexible and strong. It also aids in the body's production of natural lubrication which aids in vaginal health and better sex. Without the estrogen, the vaginal tissue becomes dry and thin and can tear during sex which is painful.

Keeping all that in mind... having sex after hysterectomy can be similar to trying to take a stroll through a landmine zone. It may be something you need to plan rather than just do spontaneously like you used to... and I recommend that... but it doesn't have to stay that way.

One of the biggest challenges is not being *allowed* to have sex until the all-clear which, for me, was scheduled at eight weeks post-op. It won't be any sooner than six weeks. I didn't understand why I was feeling the way I was. I was pushing my partner away and at first I thought maybe it was because I wasn't allowed to be intimate and I didn't want us to go too far. I felt like I was at sea. Once my hormone levels were checked it made perfect sense that I was pushing him away because my hormones were non-existent and his touching me had no emotional impact.

Once I got the okay from my surgeon I knew I'd have to try. I knew I had to try because it was important to me and it was important to our relationship and I encourage you to do the same. We were successful in making love before I had my blood test done and it wasn't bad but lubricant was definitely required. Once I got on the estrogen at nine weeks post-op it started making a difference almost right away and I found my interest, and my butterflies, returned.

What I encourage you to do is talk to your partner. Discuss this *before* surgery. Be open and honest about the surgery and the potential of this happening. Let them read this book (and as I approached writing this section I realized I should have a chapter for partners as well), or this chapter. Do some research together on the changes that *could* happen *if* you lose your ovaries or *when,* if it's planned. It is easy for people to say "oh you kept an ovary, you should be okay" but that's not what happened to me and I'm sure I'm not the only one.

You're not going to stop loving your partner, you *know* you love him but it's like you feel it differently… it feels different when there are no hormones to light the fire under the emotions. It's very unusual and disconcerting!

Once you get the okay from your surgeon, make a plan. Come up with a time that the two of you are willing to have a romantic evening. You may have to force yourself to do it. Be open and honest with each other. After six or eight weeks it almost feels foreign being touched again. And take

your time. Go slow. There's no need to rush anything.
Maybe the first time will merely be caressing and becoming
reacquainted with each other's bodies.

You will also need to give some directions. You won't
want your abdominal incision touched and you may feel
nervous about the inside of your vagina being touched. If
you've never been vocal, this is the time for you to speak up.
Let your partner know what feels good and what doesn't.
Also let him know what feels different and how. Maybe the
way he used to touch you feels worse and maybe it now feels
awesome. It's going to change over time as you become
more comfortable. After waiting for almost two months it's a
little bit like you're rediscovering each other and sex. It's fun
but scary at the same time because your body has been
abused and has new flaws plus it has changed, maybe just a
bit, but it's changed.

I encourage you to go to the store together and find a
lubricant that is acceptable to both of you. It doesn't have to
be warming or cooling or flavoured. Pick something
practical that has only a few ingredients so you don't risk
having an allergic reaction. You don't need to use a lot so it's
going to last a long time.

The next time you have sex it will be easier, and it
will get easier each time after. It gets more comfortable and
more enjoyable.

Once you start on estrogen, if it's required, you will,
again, notice that things are different. Things improve
because you will feel more emotion which will make you feel
like participating; you'll feel like you want to be there; you'll
feel more like yourself!

I have read and heard repeatedly that sex after
hysterectomy is more enjoyable. This can be due to being

Pamela Clayfield

more carefree because there's no concern about pregnancy and it can be due to no longer having to worry about periods. You're now period-free so there is no week that you feel disgusting both internally and externally.

I can't say it will be the same for everyone but my orgasms are the same now as they were which brought me much relief. In fact, I am actually able to reach a big orgasm almost every time we make love now when I wasn't able to do that before. The first time it happened, I had tears.

Talk to your doctor. Gynecologists are trained for this and are there to try to work with you to solve these problems.

Lastly, I must add, that if you're not in a relationship, and depending on your age, be sure to tell a potential new partner that you've had a hysterectomy. If you're still in a childbearing age group, a new partner might want to know right away that you are unable to have a child. Many people in their 40s are in second relationships and having children so it's important to share this with someone who may think he could still father a child.

Chapter 15... The

Hubby/Partner/Significant Other

Chapter

Hello Partner!

My apologies for the title of this chapter. I was tossing around a number of ideas for a title when I realized how many titles we have for the person we share our life with. You could be a husband or a wife, you could be a common-law partner, you could be a boyfriend and you could be a girlfriend. You could introduce each other as significant other, partner or life partner, at parties or when talking about your other half to acquaintances.

It was incredibly important to include you because, even though you are probably on board with the surgery no matter the reason it's being done, and you plan on being there and being supportive, there will be times when you

Pamela Clayfield

will feel helpless and you will start to, at the very least, dislike the lack of intimacy.

I know this because I watched my partner as I pushed him away. I didn't do it intentionally. Part of me was afraid that his hands would end up where they weren't allowed but then I realized that I was pushing him away because I wasn't feeling anything. For 15 years I had always wanted him to touch me and during my post-op time I felt broken. I felt like I had no feelings. I felt dead. My brain was desperate for the physical contact; for him to put his arms around me and hold me and tell me everything was going to be okay but my body wouldn't respond.

I encourage you to also read Chapter 14 because that is my post-hysterectomy message to the patient... your partner. Read the whole book! I'm addressing you personally in this chapter because I have the pleasure of being in an open and honest relationship and we talked, endlessly, through what I was experiencing and still do discuss it. We still sit and wish that we could go back and that I could have been able to better describe what I was feeling and he was better able to ask the right questions to try to get those answers.

I worried that I would never be interested in sex and intimacy again and I wondered if it was something I had done. I worried what would happen to us if that desire never returned. Because of how unconventional our relationship is I know he was already worried about how I would carry on in life after he was no longer in it. How does a woman build a new relationship with a new partner when she has no interest in sex and intimacy?

Thankfully it was a hormone deficiency that was easily corrected with a prescription for estrogen.

Hysterectomies & You

My advice to you is to just be there. It's not easy but this is part of relationships… the hard parts are just as important, maybe more so, than the easy times.

This will pass but it takes time and patience.

Be open with your partner about how you're feeling and point out to her what you're seeing. She may not be entirely aware and, from my experience, she may not understand why she's feeling the way she is at all. She may not even be aware of how she's behaving. I didn't realize how much I was pushing my partner away until he asked me if I wanted to see him anymore and finally told me.

There are few words to describe this lack of feelings. You feel lost, you feel numb. You are trying but just missing the mark in being affectionate because you don't desire it.

Before she sees the surgeon encourage her to talk to him about what she's *not* feeling. She's probably having hot flashes and night sweats too if it's hormone related. Some women can't discuss these things and, especially if sex was a big part of your relationship, it needs to be addressed. If sex wasn't a big part of your relationship anymore before the surgery at least encourage her to discuss the issues with the surgeon even if it doesn't change your relationship.

When she does get the go-ahead from the surgeon, go easy. Don't push. Let her take the lead but be encouraging. If there are some hormonal imbalances, tell her you want to try. Tell her you want to be intimate but that you're willing to go as far as she wants you to. Encourage her to talk to you about how she's feeling and help her find the words. If you're concerned about the potential of her withdrawing completely due to fear, go out yourself and get some lubricant. Be prepared. Encourage her to try.

Pamela Clayfield

Make a date. Set aside the time to try. Being rushed will not help. It takes longer for the body to respond and I discovered that my body responds more to his touch than it did before.

Again, we're all different. Each of us will have different surgeries; different recoveries just like each of us will have different pain experiences and healing time. Some will have an ovary removed and the other will pick up the slack and some will have the remaining ovary die like mine did.

And there may be many emotions that your partner is working through. You may find her in tears because she realizes what she has lost and that she can no longer have a baby. Your partner could feel that she isn't a woman anymore because the parts she was born with have been taken from her. Alternatively, she may be fine with the decision she made and with the loss of her uterus. I know it was a decision I made and I asked for the discussion on hysterectomy to remove the threat of a future cervical cancer because of the recurrence rate. I questioned myself once, maybe twice but I don't have a single regret. Maybe it's my training and knowledge that made me see it practically and maybe I was just determined to not fall victim to the 'C' word.

Unfortunately with surgery, the outcome… the response and the recovery is different for everyone. Your partner may have had the surgery and now be facing chemotherapy or radiation therapy or both. That means your support is necessary even longer and the idea of intimacy is wistful and a long way off.

Be present. Be there for your partner and be *with* her.

Chapter 16... Last words

No matter the reason for your surgery take it from me, you made the right choice. The decision you make is personal to you and you can't let anyone tell you otherwise. I haven't had one person tell me I shouldn't have done what I did... no, I had one person say she would never have a hysterectomy.

It's important to realize and understand this before having the surgery as it will reduce the chance of suffering from depression after the surgery. Once the surgery is done there's no turning back. In most instances, cancelling the surgery is not an option. Talk to your partner, talk to your doctor. Seek out others who have had a hysterectomy and ask them questions. Most women have little to no issue discussing their experience just like when they compare notes on childbirth. Seek out a few sessions of counseling if you feel it's necessary.

Ask questions. Talk to people who have had the surgery before you. Talk to your doctor again. Your family

Pamela Clayfield

doctor should be able to easily answer some of your questions.

If you have cancer, I wish you the best outcome of all… complete remission. In fact I hope the surgery gets every last cell so you don't require any further treatment and you can have a long, fulfilling life.

Take the time recommended and more if you can. It will make a world of difference if you take the time you need to rest and to heal. To rush back to work or do things you're not supposed to do can and likely will cause nothing but setbacks. I remember trying to move a small piece of furniture that was on rollers and ended up accidently picking up the corner of it because the wheel was caught on a carpet and I started bleeding for three days. It was something stupid that wasn't supposed to require that kind of strain, it was a freak thing, but I paid for it and that was scary.

Awareness is key and knowledge is power… be aware of what your body is trying to tell you and learn all you can about what you're having done and the rationale for the recovery time. Just knowing can calm the mind which allows you to rest more easily.

I wish you the best of luck with your surgery and hope that this book has been a wealth of information that has put your mind at ease both going into and following your surgery.

Acknowledgments

I would like to say a very special thank you to a number of my patients, friends and acquaintances who took time out to contribute. It is greatly appreciated.

I couldn't have done it otherwise.

Thank You.

Disclaimer

If you have any specific questions about any medical matter you should consult your doctor or other professional healthcare provider and not rely, solely, on the information provided in this book.

If you think you may be suffering from any medical condition you should seek immediate medical attention and never delay seeking medical advice, disregard medical advice, or discontinue medical treatment because of information in this book.

About the Author

Pamela is the author of ten novels as well as a number of non-fiction articles on health. Besides writing, she enjoys swimming, reading and movies. She has been writing for many years, has a BA in Creative Writing and also teaches writing to anyone interested in learning the craft. She lives in Waterloo, ON, Canada with her daughter.

For more information, visit her on the web at www.pamelaclayfield.webs.com or www.lulu.com/worksbyplc

Pamela Clayfield

Novels by Pamela Clayfield:

To Love Again

At Sunset

Til We Meet Again

Let The Dream Begin

The Journey Home

Mystery In The Attic

The Writing On The Wall

Changes In Time

The Trinket Box

Confessions in the Mural

www.ingramcontent.com/pod-product-compliance
Lightning Source LLC
Chambersburg PA
CBHW071228170526
45165CB00003B/1033